AI

MINDSHIFT

Unleash the Power of AI, Avoid the Pitfalls, and Keep the Human Experience

Ford Saeks

PRIMECONCEPTS
PRESS

A DIVISION OF PRIME CONCEPTS GROUP, INC.

Library of Congress Cataloging-in-Publication Data

Saeks, Ford
AI MINDSHIFT: Unleash the Power of AI, Avoid the Pitfalls, and Keep the Human Experience / by Ford Saeks.
— 1st ed.

ISBN: 978-1-884667-95-4 - Paperback

ISBN: 978-1-884667-47-3 – Hardcover

ISBN: 978-1-884667-48-0 – eBook/Kindle

1. Artificial intelligence—Business applications.

2. Strategic planning—Technological innovations.

3. Business ethics—Data protection.

4. Emotional intelligence—Artificial intelligence.

5. Leadership—Technological innovations.

Library of Congress Control Number: 2024921344

Printed in the United States of America.

Published by:
Prime Concepts Group Press, a Division of Prime Concepts Group Inc.
2637 N Shefford St, Wichita, KS 67205 USA
www.ProfitRichResults.com

First Edition, 2024, in paperback, hardcover, and e-book formats. Subsequent editions may be updated to reflect evolving technologies or new insights.

Prime Concepts Group Press publishes in a variety of print and electronic formats and by print-on-demand. Some material included in standard print versions of this book may not be included in e-books, audio, or print-on-demand editions.

Trademarks:

This book, *AI MINDSHIFT*, and its companion guide are intended for informational and educational purposes only. Any use of trademarks and logos does not imply endorsement by the respective companies, nor is any affiliation or association claimed. All trademarks and logos used in this book are the property of their respective owners, and their use is governed by fair use doctrine under U.S. trademark law. If you believe your trademark is being improperly used, please contact us to address any concerns.

Special Thanks:

Aliesa George, Devan Horning, Crystal Andrus, Linda Conner, and my E.O. Forum Members for their ongoing support, encouragement, and valuable insights. I am also deeply grateful for the trust and engagement of the many audiences from my keynote presentations, whose feedback has played a crucial role in shaping the ideas shared in this book.

The Cover design by Devan Horning, Marketing Director, PrimeConcepts.com

For volume book purchases, media interviews, or book signings, or keynote speeches, please contact Profit Rich Results at ProfitRichResults.com or call 1-316-844-0235.

Disclaimer:

This book is provided for educational and informational purposes only and does not constitute legal, financial, or business advice. Readers should consult with their own legal, financial, or business advisors before implementing any strategies or practices discussed in the book. The author and publisher disclaim any liability arising directly or indirectly from the use of the information contained herein.

Accuracy of Content

The content of this book is based on the author's personal knowledge, research, and experience as of the date of publication. While efforts have been made to ensure the accuracy of the information provided, no representation or warranty is made regarding the completeness or accuracy of the content.

Other popular titles by Ford Saeks:

For readers looking to dive deeper into business growth and leadership, don't miss Ford Saeks' other popular titles. *SUPERPOWER: A Superhero's Guide to Leadership, Business, and Life* is a must-read for anyone looking to unlock their potential and lead with confidence.

For franchisors and franchisees, check out *ACCELERATE: The Ultimate Guide for Franchisees to Maximize Local Marketing and Boost Sales.* This book offers practical strategies to drive local franchisee success.

Both books provide actionable insights that perfectly complement the AI strategies in *AI MINDSHIFT*.

Get your copies today on Amazon, or your favorite book retailer.

Download the *AI MINDSHIFT* Companion Toolkit

To dive deeper into the concepts discussed in *AI MINDSHIFT*, readers can access the companion guide for AI tutorial training videos, worksheets, and an updated list of best-practices.

Get instant access at **www.ProfitRichResults.com/aibooktoolkit**

Praise for
AI Mindshift

"Finally! This is the complete manual for using AI to meaningfully improve your business outcomes. And even more importantly, *AI Mindshift* shows you precisely WHY you need to do so."

— Jay Baer, Best-selling author of 7 books, including *The Time to Win: How to Exceed Customers' Need for Speed*

"With so much AI hype out there, *AI MINDSHIFT* stands out because it's grounded in real-world application. Ford's approach is pragmatic, and his insights are spot-on for anyone looking to adopt AI without losing sight of what makes their business human. Our team has already started applying his ideas, and the results are tangible."

— Uri Geva, CEO Cookie Dough Bliss Franchisor

"Ford's approach to AI is refreshing. In *AI MINDSHIFT*, he shows how to leverage AI for better decision-making and productivity, while never losing the human touch that sets businesses apart. The insights in this book are timeless and immediately actionable."

— Craig Merrills, Franchise Partner WOW 1 DAY PAINTING

"Ford Saeks has a way of making complex topics like AI accessible and actionable. *AI MINDSHIFT* is a game-changer for leaders who want to innovate while staying connected to the human side of business. This book isn't just about technology; it's about using AI to create real value, without losing what makes us unique."
— Aliesa George, CEO HZC Enterprises & Centerworks.com

"Ford Saeks has nailed it with *AI MINDSHIFT*. As a serial entrepreneur, I've seen a lot of hype around AI, but Ford's approach is different. It's a must-read for anyone serious about leveraging AI for business growth."
— Will Ezell, CEO, Biz Visioneers

"*AI MINDSHIFT* changed the way we think about AI. Ford Saeks shows how to implement AI without losing sight of the human experience, which is so critical for maintaining customer loyalty."
— Shep Hyken, Customer Service/CX Expert And New York Times Best Selling Author

"Ford Saeks has the unique ability to simply the complexity of AI in a way that mere mortals understand and can apply AI to their personal and professional lives. AI MINDSHIFT is a roadmap to using AI strategically to keep your team and customers engaged. No matter what your experience with AI, Ford's book will provide you with invaluable insights."
—Patricia Fripp, President, A Speaker For All Reasons

"In a world obsessed with AI, AI MINDSHIFT is the strategic wake-up call every organization needs. Ford Saeks reveals how to harness AI's power without losing the human connection—essential reading for those who want a competitive edge."
— Daniel Burrus, New York Times Bestselling Author, Leading Technology

"*AI MINDSHIFT* was a breath of fresh air. Ford Saeks lays out a clear, no-nonsense roadmap for AI adoption that puts people first. His focus on maintaining trust, empathy, and customer relationships is exactly what today's AI conversations are missing."
— Maureen Zappala , CEO High Altitude Strategies

"Ford Saeks' *AI MINDSHIFT* gives a clear path to leveraging AI without falling into the traps of over-reliance and automation overload. It's a thoughtful, strategic guide that shows how to implement AI in ways that enhance both business performance and customer relationships."
— Katrina Mitchell, CEO & Chief Match Maker, Franchise Speakers

"As a leader in a rapidly growing company, I was overwhelmed with how to approach AI until I read *AI MINDSHIFT*. Ford shows exactly how to balance technology with human intuition. It's not about automating everything—it's about being smart with AI to amplify what already works."
— Art Leger, CEO Clean Rite Pressure Washing

"Ford's *AI MINDSHIFT* helped us see AI not as a replacement for our team but as a partner in driving growth. He gets that AI is about enhancing human potential, not replacing it. This book gave us the blueprint to implement AI in a way that's authentic and results-driven. Highly recommended for any one navigating AI adoption."
— Joe Prenatt, Founder, Pressure Washer Sales & Service LLC

"Ford Saeks does an exceptional job of showing how AI can transform business without compromising the human element. *AI MINDSHIFT* is filled with insights that are not only thought-provoking but also immediately actionable. It's the best guide I've read on making AI work for businesses without losing authenticity."
— Tim Gard, CSP, Hall of Fame Keynote Speaker

"In a sea of tech jargon and over-hyped AI promises, Ford Saeks offers a refreshing, practical perspective. *AI MINDSHIFT* is about integrating AI with purpose, and it's helped me see how we can use AI as a tool for growth while staying true to our core values. This book is a valuable resource for any leader serious about leveraging AI."
— Scott McKain, Founder and CEO, The Distinction Institute

Table of Contents

Introduction 1

AI Mindshift Book Toolkit 6

Section One: Critical Thinking in the Age of AI

Chapter 1: How to Get the Most Out of This Book 9

Chapter 2: Rethinking Business in the Age of AI 21

Chapter 3: AI Prompting: Beyond Simple Queries.......................31

Chapter 4: The Human-AI Balance - Maintaining High Touch ...39

Section Two: Implementing AI Across Business Functions

Chapter 5: AI Mastery: The Four Stages of Seamless Integration .51

Chapter 6: AI for Leadership and Decision-Making 61

Chapter 7: AI and Emotional Intelligence (EQ) – A Winning

Combination 69

Chapter 8: Organizational AI Adoption Strategies.......................77

Section Three: Navigating Pitfalls and Avoiding Common AI Challenges

Chapter 9: Common AI Pitfalls and How to Avoid Them 87

Chapter 10: Navigating Data Privacy, Bias, and Fairness –
Creating Ethical AI Practices 97

Chapter 11: AI in Crisis Management – When Speed Matters 109

Chapter 12: Navigating the AI Compliance Maze 119

Section Four: Driving Innovation and Human-Centered AI for Lasting Success

Chapter 13: Why Speed of Iteration Matters in AI 127

Chapter 14: Choosing the Right AI Tools for Your Business 137

Chapter 15: Continuous Improvement and AI Innovation 147

Chapter 16: Moving Forward– Maintaining the Human Touch
in a Tech-Driven Future 157

Chapter 17: The AI Horizon: Emerging Trends and Their
Business Impact 167

Chapter 18: AI Mindshift Action Plan – Your Next Steps...........177

Recommended Resources 185

Keynote Presentations and Training Solutions. 185

60 Ways to Use AI in your Business and Prompting Guide.....193

About the Author...211

Introduction

AI isn't just changing the way we do business—it's revolutionizing how we think about business. And here's the truth: adopting AI isn't about using the newest tools or keeping up with trends. It's about making smarter decisions, faster. It's about blending human insight with machine intelligence to unlock opportunities that were impossible just a few years ago.

But let's cut through the hype. Everywhere you look, people are talking about AI like it's the magic bullet that will solve all your problems. The reality is, AI won't fix what's broken in your business. *How you think about AI will.*

That's where the AI Mindshift comes in. This book isn't about turning you into an AI expert. It's about helping you rewire how you approach AI—using it as a strategic asset, not just another shiny object to throw at your problems. The leaders who thrive in this new landscape aren't the ones with the fanciest tech. They're the ones who think critically, ask the right questions, and use AI to amplify what makes their business unique.

The Real AI Advantage: Thinking Differently

AI is everywhere, and if you're not using it strategically, you're already falling behind. The game has changed, and business as usual won't cut it anymore. But here's the kicker—most companies are implementing AI the wrong way. They're automating tasks,

1

crunching data, and replacing processes without truly understanding how to make AI work for them.

They're missing the big picture. They're missing the Mindshift.

You're not here to be like everyone else. You're here because you want to stay ahead of the curve. You want to think differently, act strategically, and lead in a way that makes AI work *for* your business—without losing what makes it human.

Did I Use AI to Write This Book?

Let's tackle the elephant in the room: **Did I use AI to write this book?** The short answer is yes—but not in the way you might think.

Here's the reality: AI is a tool, and like any tool, its value depends on how you use it. In the process of writing this book, I used AI strategically—to organize research, refine ideas, and explore multiple perspectives quickly. It helped me accelerate certain parts of the writing process, just like AI helps business leaders automate and optimize specific tasks.

But here's the key point: **AI didn't write this book for me.** It didn't generate the insights, the stories, or the frameworks that make this book valuable. Those come from decades of experience, working with businesses of all sizes, helping leaders navigate the real-world challenges of growth and innovation.

AI may have enhanced the process, but the human element is irreplaceable. The leadership strategies, critical thinking frameworks, and actionable insights in this book come from my experience, intuition, and deep understanding of what makes businesses succeed. AI can amplify what's already there, but it can't replace the strategic thinking and expertise that guide every page.

That's exactly what this book is about—understanding how to leverage AI to *amplify* your strengths, not replace them.

It's Not About AI, It's About Execution

Let me be blunt: AI is a tool, and like any tool, its value depends on how you use it. I've spent decades working with businesses of all sizes, from startups to billion-dollar franchises, and here's what I've learned: It's not what you know—it's how well you execute.

You can have the best AI tools in the world, but if you don't understand how to implement them, you're just going to waste time, money, and energy. That's why this book focuses on practical strategies you can put into action right away.

AI isn't going to replace you. It's not going to replace your team, your relationships, or the human connections that drive real business growth. But it *will* change the way you work, lead, and compete. The question is: *Will you embrace it and thrive, or will you get left behind?*

Developing an AI-Critical Thinking Mindset

This isn't about diving into the technical weeds or learning every nuance of machine learning. It's about developing an AI-critical thinking mindset—a way of approaching AI that helps you make smarter decisions, avoid common pitfalls, and leverage AI in a way that's strategic and sustainable.

Ask yourself:

- What problem am I really trying to solve with AI?

- Where does AI fit into my business strategy?

- How can AI enhance my team's strengths, rather than replace them?

- What are the ethical and human impacts of adopting AI in my organization?

These are the questions that separate the leaders who thrive from those who struggle. The real value of AI isn't in automating everything or removing human intuition from decision-making. It's in using AI as a powerful force multiplier—allowing you to innovate faster, think more critically, and make better choices in real time.

The Stakes Are High—and the Opportunity Is Bigger Than Ever

Let me be clear: **AI is here to stay.** It's reshaping industries, changing customer expectations, and forcing leaders to rethink everything they know about business. If you don't adapt, you'll be left behind. But if you can master this mindshift, you'll position yourself—and your business—for long-term success in a world where AI is only going to become more integrated into everything we do.

What's Ahead: Your AI Mindshift Journey

In the coming chapters, we're going to break down exactly how to develop this AI-critical thinking mindset and use it to drive results. **You'll discover how to:**

- **Think critically** about AI, asking the right questions and making data-driven decisions that align with your business goals.

- **Leverage AI across your organization,** from leadership and decision-making to customer experience and operational efficiency.

- **Avoid the common AI pitfalls** that are derailing so many businesses.

- **Drive innovation,** without sacrificing the human touch that sets your business apart.

This book isn't about giving you a crash course in AI technology. That would be fruitless because the technology is changing daily. It's about helping you make AI work for *you*—so you can stay competitive, relevant, and ready for the future.

Let's get started.

AI Mindshift
Book Toolkit

Unlock Exclusive AI Tools & Resources with Your Purchase!

Congratulations on taking the next step to accelerate your business with AI! To help you get even more value from your purchase of *AI MINDSHIFT*, I'm offering **exclusive access** to my AI Book Toolkit—designed to provide you with the essential tools, templates, and strategies to implement AI in your business right away.

Here's What You'll Get:

- **Bonus Video Tutorials**: Learn directly from Ford Saeks with actionable tips for maximizing your AI efforts.

- **AI Prompt Templates**: Ready-to-use prompts that can generate high-quality content, marketing strategies, and sales solutions.

- **Step-by-Step Action Plans**: Follow easy-to-implement AI strategies that will help you streamline tasks and boost productivity.

- **Early Access to AI Webinars**: Be the first to know about upcoming live training sessions and interactive webinars.

Get Instant Access to the Bonuses:

All of this is available exclusively for readers of *AI MINDSHIFT*.

Scan the QR code or simply visit:
ProfitRichResults.com/aibooktoolkit to unlock these powerful resources and start applying AI to your business today!

Section One:
Critical Thinking in the Age of AI

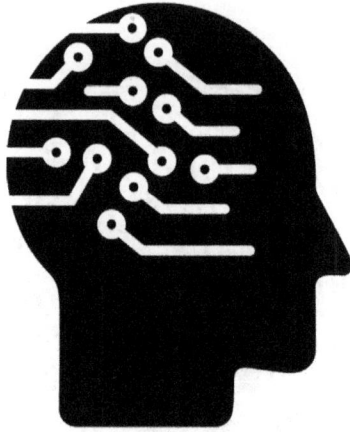

AI is more than just a technological shift—it's a game-changer for how we think, operate, and compete. But here's the kicker: mastering AI starts in your mind. This section is all about rewiring how you approach AI, making sure you don't fall into the trap of thinking AI is just another tool. It's not. It's the future of business strategy, and those who understand how to think critically about AI will thrive. Get ready to sharpen your mental edge and start looking at AI as a strategic partner, not just a shiny new gadget.

Chapter 1:

How to Get the Most Out of This Book

Welcome to Your AI Mindshift Journey

AI is here, and it's not waiting for anyone. Whether you realize it or not, and I'm sure you do, AI is already reshaping the business landscape, and if you're not thinking critically about it, you're already falling behind. It's not just a buzzword—it's the driving force behind the future of work, decision-making, and growth.

But here's the key: **AI by itself isn't enough.** It's not just about plugging in new tools or automating a few tasks. The real competitive edge comes from how you think about AI—how you integrate it into your business strategy and use it to make better, faster decisions. That's what this book is about: **changing how you think** about AI, so you can harness its power, stay relevant, and avoid being left in the dust.

Who Is This Book For?

If you work in business, this book is for you. Whether you're a CEO, manager, team leader, entrepreneur, or even someone just starting their career—if you have a stake in how your company runs, you need to understand AI and how it's going to change the game. This book isn't written for tech gurus or data scientists. It's written for people like you, who are responsible for getting results and making decisions that impact your team, your customers, and your bottom line.

You might be leading a team, managing operations, running a department, or handling customer service. You could be in sales, marketing, HR, finance, or product development.

No matter your role, AI is going to affect how you do your job—and this book is designed to help you not only adapt but thrive.

- **For business owners**: You'll discover how to use AI to streamline processes, reduce costs, and boost your competitive edge.

- **For managers and department heads**: You'll learn how to integrate AI into your team's workflow to improve efficiency, without sacrificing the human touch that drives engagement.

- **For entrepreneurs**: You'll see how AI can give you the agility to compete with bigger players, helping you punch above your weight.

- **For employees at any level**: You'll find ways to future-proof your career by understanding how to work with AI, not fear it.

In short: If you're involved in business at any level, this book is for you. AI is coming for every industry, every department, and every role. The question is, will you be ready?

How to Get the Most Out of This Book

Let's cut straight to it: **This book is about critical thinking and action, not theory.** I'm not here to give you a lecture on AI's inner workings or to impress you with tech jargon. You don't need to be a programmer or a data scientist to benefit from AI. What you need is a clear strategy on how to apply AI to real-world business challenges, and that's exactly what you'll get.

Keep These Steps in Mind:

1. **Start with your mindset**: The first thing you need to under-stand is that AI is not just another tool. It's a shift in how you approach problem-solving, decision-making, and lead-ership. Throughout this book, I'll challenge you to rethink how you operate, using AI as a multiplier for your existing strengths.

2. **Dive into the MINDSHIFT EXERCISES**: At the end of every chapter, you'll find practical, hands-on exercises to help you apply what you've learned. These **MINDSHIFT EXERCISES** aren't just homework—they're designed to push you into action, get you thinking differently, and drive immediate results in your business.

3. **Leverage the companion resources**: The world of AI is evolving fast, and so is this book. Alongside the content in these pages, I've built a **living toolkit** at ProfitRichResults. com/aibooktoolkit, (see page 6) where you'll find updated templates, videos, and additional resources. As AI advances, you'll have access to the latest strategies to stay on top of your game. It's included with the purchase of this book and you get instant access to all the tools online.

4. **Look for the quick wins**: AI doesn't have to be overwhelm-ing. Throughout this book, I'll point out **Quick Wins**—small, high-impact actions you can take immediately to start seeing results. You don't need to completely overhaul your business overnight. Start small, build momentum, and let AI work for you.

Why This Book Is Relevant—Even in a World of Rapid Change

By now, how many times have you hear on the news or countless spam emails with the next AI webinar: **AI is changing everything.** And yes, the technology is evolving at a breakneck pace. But here's why this book is different—and why it will remain relevant, even as new tools and technologies emerge.

This book is about strategy, not just technology. The tools will change, sure. But the way you think about AI—how you use it to improve decision-making, streamline operations, and drive innovation—will always be the differentiator. This book will give you the frameworks and mindsets you need to stay ahead, no matter what new AI tool hits the market next.

The danger of doing nothing is real. Sitting back and waiting for AI to "settle down" isn't an option. Every day you hesitate, your competition is using AI to get faster, smarter, and more efficient. That's why it's so important to get started now. You don't need to master every tool—you need to master the mindset. Once you have that, you'll be ready to adapt to whatever comes next.

Why Me? Why This Book?

So why should you listen to me? **Because I've been in the trenches**—helping businesses navigate uncertainty, scale faster, and dominate their markets. Over the last few decades, I've worked with everyone from startups to billion-dollar franchises. I've seen first-hand what works and what doesn't when it comes to implementing new technology in a way that drives real business growth.

I'm not here to impress you with AI theory. I'm here to give you the tools to get results. I've built companies, scaled businesses, and consulted with leaders who were skeptical about AI—until they saw the real-world impact. I'm not an AI engineer, and that's exactly why you should listen. Because like you, I'm a business leader who cares about results, not tech fads.

This book is built on real experience—working with businesses to leverage AI in practical, meaningful ways. I've been where you are, and I know what it takes to drive growth. AI is a powerful tool, but only if you know how to use it strategically. As I said to skeptics in the past, "Just google my name and check out my many 5 star verified Google reviews"

Why You Can't Afford to Ignore This

Let me put it bluntly: **The stakes couldn't be higher.** AI is reshaping industries faster than most businesses can keep up, and those that fail to adopt an AI-driven strategy are going to get left behind. This isn't a "nice-to-have" anymore—it's a business imperative.

Every day that you delay is a day that your competitors are gaining ground, automating processes, making data-driven decisions, and delivering faster, smarter results. If you wait until it's comfortable, you'll already be too late.

But here's the good news: **You don't need to master AI overnight**. You just need to start. This book will give you the frameworks, strategies, and mindset shifts you need to get ahead. The rest will come. But make no mistake—doing nothing is the biggest risk of all.

Ready to Make the Shift?

This isn't just another business book. It's your guide to navigating the future. AI is changing everything, and those who embrace it with the right mindset will thrive. If you're ready to think critically, take action, and future-proof your business, this book will show you how.

We're at a turning point, and the choices you make today will define your business tomorrow. The greatest fear with AI isn't adopting it too soon, it's waiting to long and falling behind.

Quick Wins:

❖ **Identify one area in your business where AI could help immediately**—whether it's improving efficiency, streamlining processes, or enhancing customer experience. Don't over complicate it. Start with a small, manageable task.

❖ **Set up a meeting with your team** to discuss AI and how it could impact their specific roles. Begin by asking each team member where they see opportunities for automation, smarter decision-making, or AI-driven innovation.

❖ **Bookmark the AI Toolkit at** ProfitRichResults.com/aibooktoolkit. As you read through the book, check back regularly for updated resources, templates, and tips on how to implement the strategies you're learning. Okay, I've mentioned that 5 times, so go link above or to page 6, scan the QR code and get the bonuses.

Key Takeaways:

❖ **AI is no longer optional.** It's already transforming industries, and those who don't adopt an AI-driven mindset are at risk of falling behind.

❖ **This book is designed for action.** It's not just about understanding AI; it's about making smarter decisions, faster, and using AI to amplify your business strategy.

❖ **AI is a tool—your mindset is the differentiator.** It's not about the latest tools and technologies, but how you think about and apply AI to solve real business challenges.

❖ **Success comes from taking the first step.** Don't wait for AI to settle down or become "easy." Start now by developing the right mindset, and you'll be prepared for whatever comes next.

MINDSHIFT EXERCISE #1:
Shifting Your AI Mindset

Take 15 minutes to reflect on how AI is—or isn't—currently impacting your business. Use the following prompts to start thinking critically about where AI could make a difference:

❖ **Where do you see AI being used in your industry already?** Identify at least three ways companies in your space are leveraging AI, even if you aren't yet.

❖ **What's one area of your business that feels inefficient or repetitive?** Think about tasks, processes, or decisions that could benefit from automation or data-driven insights. Write down a few ideas on how AI might help improve them.

❖ **Where could AI enhance—not replace—your team's strengths?** Reflect on how AI can work alongside your people, making them more effective rather than redundant. Consider tasks where AI could free up time for your team to focus on higher-value work.

❖ **What's your biggest fear about AI?**
Acknowledge any concerns you have
about adopting AI—whether it's a
fear of losing control, complexity, or
replacing the human element. How can
you address this fear and turn it into an
opportunity?

Action Step: Share your thoughts from this
exercise with a colleague, team member, or mentor,
and have a discussion about where AI could fit into
your business. Don't aim for perfection—just get
the conversation started.

**You've just taken the first step in shifting
your mindset toward AI, but this is just the
beginning.** The real challenge is making sure AI
works for you, your team, and your business, rather
than getting overwhelmed by the technology itself.
In the next chapter, we're going to dive deeper into
how you can reframe your entire approach to AI,
moving beyond basic automation and into strategic
integration that drives real results.

Chapter 2:

The AI Mindshift: Rethinking Business in the Age of AI

AI isn't just about automating tasks — it's about amplifying our capacity to innovate and solve complex problems.

AI is here, it's powerful, and it's not going away. But here's what you need to know right off the bat — it's not about the technology. It's about how you think about the technology.

If you've picked up this book, you're probably feeling a mix of excitement and anxiety about AI. Maybe you're worried it'll make your job obsolete. Or perhaps you're eager to use it but don't know where to start. Whatever brought you here, I'm glad you're ready to tackle this head-on.

I'm going to be straight with you: AI is reshaping our world faster than most of us can keep up. But here's the kicker — success in this new landscape isn't about who has the fanciest AI tools. It's about who can think critically about how to use them.

The Wake-Up Call

Let me paint a picture for you. I recently worked with a mid-sized manufacturing firm. They were chomping at the bit to implement AI across their entire operation. Sounds progressive, right? Wrong. They hadn't stopped to consider how it would impact their creative process, their client relationships, or their team dynamics. The result? A lot of wasted time, money, and a team that felt like they were fighting against the machines instead of working with them.

This is the reality for too many businesses right now. They're so afraid of being left behind that they're jumping into AI without a life jacket. Don't be that business.

What You Need to Know

Here's what this book is going to do for you:

1. It's going to help you think about AI, not just how to use it.

2. It's going to show you how to spot the real opportunities and the real threats.

3. It's going to give you practical, no-nonsense strategies for implementing AI in a way that actually makes sense for your business.

But most importantly, it's going to help you develop what I call an "AI-critical thinking mindset." This isn't some fancy buzzword. It's a practical approach to evaluating and using AI that will set you apart in a world where everyone else is just following the herd.

The AI-Critical Thinking Mindset

So what exactly is this mindset? It's about asking the right questions:

- What problem am I really trying to solve?

- Is AI actually the best solution for this problem?

- How will this impact my customers, my employees, and my bottom line?

- What are the ethical implications?

- How can I use AI to enhance human capabilities, not replace them?

It's about understanding that AI is a tool, not a magic wand. And like any tool, its value depends entirely on how you use it.

Shaping Your Approach to AI: A Simple Framework

Here's a framework I use with my clients to make sure AI actually helps and doesn't get in the way:

1. **Identify the Problem**: Don't start with AI—start with the challenge you're trying to solve. Is it customer engagement? Operational inefficiencies? AI should serve a purpose, not just be a shiny object.

2. **Evaluate AI's Fit**: Once you know the problem, ask if AI is really the right solution. Sometimes it is. Sometimes it's not. Be honest about where AI can add value.

3. **Balance Human and AI**: AI isn't here to replace your people—it's here to help them. Where can AI take over repetitive tasks, and where do you need humans to step in? Find the balance.

4. **Test and Adapt**: AI is constantly evolving. Your approach should be, too. Implement in stages, test the results, and make adjustments as needed.

This framework helps ensure you're not just adding AI because it's trendy, but because it's going to make a real difference.

The Stakes

Let me be clear: the stakes here are high. AI isn't just another tech trend. It's a seismic shift that's going to separate the businesses that thrive from those that barely survive. Like you've heard me say in my keynotes and podcasts... "it's not what you know, it's how well you execute!"

But here's the good news: you don't need to be a tech genius to come out on top. You just need to think critically, stay adaptable, and always keep your focus on the human element of your business.

Quick Win:
AI-Critical Thinking Exercise

Take 5 minutes right now and jot down answers to these questions:

- ❖ What's the biggest challenge in your business right now?

- ❖ How might AI potentially help with this challenge?

- ❖ What are the potential risks or downsides of using AI for this?

- ❖ How would you ensure the human touch remains if you implemented AI here?

The Path Forward

In the coming chapters, we're going to dive deep into each aspect of the AI-critical thinking mindset. Here's a glimpse of what's ahead:

- ❖ We'll explore how to have meaningful conversations with AI, moving beyond simple queries to true AI collaboration.

- ❖ You'll discover how to maintain the human touch while integrating AI across various business functions.

❖ We'll tackle common AI pitfalls and challenges, including ethical considerations and crisis management.

❖ Finally, we'll look at driving innovation and ensuring lasting success with AI, always keeping the focus on human-centered solutions.

Throughout this journey, you'll encounter real-world examples, practical strategies, and actionable insights you can implement right away. From leadership and decision-making to emotional intelligence and compliance, we'll cover the full spectrum of AI's impact on business.

But remember, this journey starts with a shift in thinking. It's not about the technology — it's about how you think about the technology. By changing your approach to AI, you're positioning yourself and your business to leverage its full potential, turning challenges into opportunities for growth and innovation.

Key Takeaways

- ❖ AI is reshaping business, but success depends on how you think about it, not just how you use it.

- ❖ Developing an AI-critical thinking mindset is crucial for navigating this new landscape.

- ❖ Always start with your business problem, not the technology.

- ❖ Balance AI capabilities with human skills and judgment.

- ❖ Stay adaptable and always keep the focus on enhancing the human experience, not replacing it.

MINDSHIFT EXERCISE #2:
The AI-Critical Thinking Journal

Take 10 minutes to reflect on AI's current impact in your business. Answer these questions:

- ❖ What's the biggest challenge or opportunity you're facing today?
- ❖ How could AI help you solve this challenge or amplify this opportunity?
- ❖ What risks could emerge from using AI in this area?
- ❖ How will you balance AI's role with the human strengths of your team?

Action Step: Begin an "AI-Critical Thinking Journal" to regularly reflect on AI's role in your business and refine your approach as technology evolves.

Now that you've grasped the mindset needed to navigate AI's transformative potential, it's time to move beyond just knowing how AI works. In the next chapter, you'll discover how to communicate with AI effectively by mastering the art of prompting—unlocking deeper insights that can give your business a competitive edge.

Chapter 3:

AI Prompting: Beyond Simple Queries

Mastering AI prompts is like learning a new language — one that speaks directly to the future of your business.

AI isn't about getting quick answers—it's about opening up a dialogue that drives deeper insights and practical strategies. Think of AI like the best consultant you've ever hired—it's ready to provide value, but only if you ask the right questions.

We're all accustomed to search engines like Google: you type a question, and boom, instant results. But AI? That's a different ball game. You don't just throw in a query and walk away with a single answer. You're having a conversation. And the better you get at prompting AI, the more useful it becomes.

Moving from Search Queries to Conversations

Here's a shift in mindset: Instead of treating AI like a static search tool, think of it as an interactive guide. Asking it generic questions like "What's new in marketing?" will get you broad answers. But a question like "How can I improve customer retention in my B2B business while personalizing our email marketing?" delivers a deeper, more valuable response.

The key difference is nuance. The better you craft your prompt, the more precise and actionable the AI's response will be. This isn't just about saving time—it's about getting tailored solutions that move the needle for your business.

Imagine this: You're sitting with an expert who has instant access to all the data in the world. Would you ask them vague questions or dig into specifics to get real insight? AI works the same way.

Best Practices for Prompting Multimodal AI Models

So how do you get better at crafting these conversations? Let's break it down:

- **Provide Context:** The more background you give, the better AI will perform. Instead of asking, "What's a good sales strategy?" say, "For a small B2B tech company focusing on recurring revenue, what's a solid sales strategy?"

- **Be Specific**: AI isn't a mind reader. If you want an in-depth report, ask for it. If you're looking for a simple summary, say that. Clear instructions result in clear answers.

- **Iterate Your Questions**: Sometimes the first response won't be what you're looking for. Treat it like a back-and-forth. Refine your prompts and ask follow-ups like "Can you expand on how this applies to small businesses?"

- **Leverage Multimodal Capabilities:** AI isn't just for text anymore. Many tools can analyze images, data sets, or even voice inputs. Take advantage of this, especially if you're in fields like marketing or design.

By following these practices, you'll move beyond surface-level answers and unlock insights that can transform your business.

Lessons in Communicating with AI-Driven Tools

Think of AI communication as a funnel. Start broad to explore possibilities, then narrow down to specifics. Ask follow-up questions, rephrase, and test different angles until you get the most valuable response.

- **Start Broad, Then Narrow**: This allows you to explore the AI's range before diving deep into specifics.

- **Leverage Follow-Up Prompts**: Don't just accept the first answer. Keep the conversation going with clarifications or re-angled questions for more focused results.

- **Test New Angles**: If you're not getting the depth you need, rephrase the prompt. Instead of asking, "What's the best marketing strategy?" ask, "What are proven marketing strategies for small businesses with limited budgets?"

The Collaboration Mindset

Treat AI as your partner in collaboration. It's not just a tool to spit out answers—it's a system that helps you refine your thinking, explore new ideas, and spot opportunities. When you approach it as a collaborative process, you'll unlock its full potential.

Quick Win: AI Conversation Starter

Take a business problem you're currently facing and write a prompt for an AI system to solve it. Start with "I need a strategy to..." Then, refine that prompt three times, each time adding more context or constraints. This will help you practice crafting effective prompts and get more precise insights.

Checklist for Crafting Effective AI Prompts:

❖ **Define the Purpose:** What problem are you trying to solve? *Example:* "I need a customer engagement strategy for a small B2B business."

❖ **Provide Context:** Who or what is this related to? *Example:* "For a small B2B company focusing on increasing recurring revenue."

❖ **Ask Specific Questions:** Make sure your questions are clear and focused. *Example:* "What are effective customer engagement strategies for a small B2B company focused on recurring revenue?"

❖ **Request a Specific Format:** How should the AI respond? *Example:* "Please provide three strategies in bullet points."

❖ **Refine and Follow Up:** Ask follow-up questions for more clarity or details. *Example:* "Can you expand on how these strategies differ for small vs. medium-sized businesses?"

❖ **Checkout the latest AI Video Training and Prompting Guide book** at: https://ProfitRichResults.com/ai-training

Key Takeaways:

❖ **Treat AI as a Conversation Partner:** Engage with AI like you're having a dialogue. The better your prompts, the better your results.

❖ **Provide Clear Context and Specific Instructions:** Tailor your prompts by giving AI the context and detail it needs to provide precise, valuable responses.

❖ **Refine Through Iteration:** Don't settle for the first answer—ask follow-up questions to drill down and improve the quality of AI's insights.

❖ **Leverage Multimodal Capabilities:** Use AI's ability to process text, images, and other formats to get richer, more comprehensive outputs.

MINDSHIFT EXERCISE #3:
The Prompt Optimization Workshop

Take a challenge you're facing in your business and write a prompt for an AI tool to assist with it. Refine the prompt three times:

- ❖ **Initial Prompt:** Write the first version of the question you're asking AI.

- ❖ **Refinement:** Add more specific goals or instructions.

- ❖ **Final Prompt:** Provide detailed context or examples for the AI to generate more tailored insights.

Action Step: Practice refining AI prompts and note how each version impacts the AI's response. Use this technique regularly to improve the quality of your AI-driven insights.

By treating AI as a collaborator rather than a search tool, you'll unlock deeper insights and drive more meaningful outcomes for your business. You've just learned how to have meaningful conversations with AI, but this is only the beginning. In the next chapter, we'll dive into how to maintain that crucial human touch while leveraging AI, so your business can stay efficient without losing its authentic human connection.

Chapter 4:

The Human-AI Balance — Maintaining High Touch

AI can scale your efficiency, but it's the human touch that scales your relationships.

In the AI era, the most valuable currency isn't data—it's empathy. AI can provide efficiencies and valuable insights, but without a human connection, you risk turning your business into just another faceless machine. True success comes when we strike a balance between speed and empathy, using AI to enhance, not replace, our relationships.

The Automated Call Tree: A Case Study in Losing the Human Touch

Remember the rise of automated call trees? Companies were promised a revolution—save time, money, and resources by automating customer service. But the reality? Customers felt frustrated as they navigated through endless voice prompts, desperate to reach a human. What started as a promising innovation quickly turned into a source of disconnection.

This is the cautionary tale for businesses integrating AI. If you prioritize automation at the expense of human connection, you risk alienating the very people you aim to serve. The lesson here is that **balance** is key. AI can handle the repetitive tasks, but the heart of your customer relationships must remain human

Finding the Sweet Spot: Blending AI and Human Interaction

So, how do you avoid the mistakes of the past? By recognizing where AI can be most helpful and where humans are indispensable.

1. **Use AI for Routine, Not Emotion**: AI excels at handling repetitive tasks—responding to FAQs, processing transactions, or updating account statuses. This can free up your team to focus on more complex, emotionally charged interactions. But remember, AI can't empathize. For those moments when a customer is upset, they need to feel heard and understood by a real person.

2. **Keep Humans at Critical Touchpoints**: Just like the best customer experiences, the most successful business processes combine speed and efficiency with empathy and connection. AI can handle the basics, but when situations escalate, humans must step in to manage higher-stakes decisions or emotionally sensitive interactions.

3. **Test and Adapt**: Like the call tree systems of old, AI solutions need constant evaluation. What works today may not work tomorrow. Continuously gather feedback and adjust your AI systems to ensure you're delivering an efficient yet empathetic customer experience.

Case Study: A Balanced Approach to AI in Customer Service (And My Own Experience)

Let's take a closer look at how blending AI and human interaction can work, using both a real-world example and my personal experience.

A mid-sized retail company implemented an AI-powered chatbot to handle customer inquiries. At first, the results were impressive: response times improved, and customer satisfaction with basic inquiries went up. However, they quickly found that when it came to more nuanced or emotionally charged situations—like processing returns or handling complaints—customers preferred speaking to a human.

They learned from this and implemented a hybrid system: the chatbot would handle simple questions and transactions, but any conversation that required empathy or deeper understanding was automatically routed to a human representative. As a result, they were able to scale their customer service operation while maintaining the personal connection that kept customers loyal.

My Own Journey with Chatbots

I've always wanted to have a chatbot on my speaking and agency websites. The idea sounded great, but the execution was daunting. Coming up with all the questions, potential answers, and branching responses felt overwhelming—and frankly, the early results weren't that helpful. The bots just didn't effectively engage visitors or move them toward taking action.

But things changed recently when I set up smart AI chatbots using my **Zapier** account. Instead of manually crafting every possible interaction, I simply uploaded relevant documents about my business, services, and products, and linked the chatbot to the right sections of my website. Now, the AI chatbot simulates human interaction, engaging visitors, answering their questions, and even having a little fun along the way.

What's more, after a few questions, the chatbot asks for their contact information—helping me generate new leads without lifting a finger. This experience showed me that when you let AI do what it's best at (handling routine interactions), it frees up your time to focus on more strategic activities.

Leveraging AI to Empower, Not Replace, Your Team

While AI can take over many repetitive tasks, the goal should never be to eliminate human involvement entirely. Instead, AI should empower your team to do their best work by freeing them from administrative tasks and giving them more time to focus on high-value activities.

- **In leadership**: AI can provide data-driven insights, but human leaders are still needed to make judgment calls, understand team dynamics, and provide emotional support.

- **In creative fields**: AI can assist with content generation or analysis, but it's the human touch that adds context, innovation, and nuance to the final product.

The Business Impact of Balancing AI and Human Touch

Why does this balance matter so much? Because in a world where automation is becoming more common, the businesses that maintain strong human connections will stand out. Think of it like this: AI is quickly becoming table stakes. Everyone will have it. But what will set you apart is how well you keep the human element alive in your customer interactions, leadership, and culture.

Let's take **leadership** as an example. AI can analyze performance data and suggest improvements, but it can't understand the underlying emotions affecting team morale. Leaders who rely too heavily on AI to make decisions risk losing touch with their teams. Conversely, leaders who use AI to gather insights while maintaining personal relationships will build stronger, more motivated teams.

How to Ensure AI Enhances Rather Than Erodes Human Connection

1. **Constant Feedback and Testing**:

   o You need to monitor how your AI systems interact with customers and employees. Regularly gather feedback to ensure AI isn't causing frustration or distancing you from your audience. Make adjustments where needed.

2. **Training Your Teams to Use AI Effectively**:

   o AI is a tool, and like any tool, it's only as effective as the people using it. Train your team to understand when and how to use AI, and when to step in to provide the human touch.

3. **Integrating Empathy into AI Systems**:

   o Even in automated interactions, it's possible to

program empathy into the system. For example, you can set your AI to recognize frustration in a customer's language and immediately offer to connect them with a human representative.

4. **Personalize Where Possible:**

- Use AI data to create personalized experiences. For instance, leverage customer history to make interactions feel tailored and meaningful, not robotic. This keeps the customer feeling recognized and respected.

5. **Avoid Over-Automation**

- Resist automating every touchpoint. Identify critical stages in the customer journey where human interaction is essential. A well-timed personal touch can build trust and loyalty that AI alone can't achieve.

Quick Win:
Human-AI Balance Audit

Take 15 minutes to audit one of your customer-facing processes. Identify where AI is currently used and where human interaction is necessary. Create a flowchart that shows these touchpoints. This simple exercise will help you optimize the balance between efficiency and empathy, improving both customer satisfaction and operational efficiency.

Key Takeaways:

❖ **Avoid Over-Automation:** Keep a careful balance—AI should handle repetitive tasks, but human involvement is needed for empathy and complex decision-making.

❖ **Use AI for Speed, Not Emotion:** AI can make processes faster, but it can't replace the emotional intelligence required in customer service, sales, and leadership.

❖ **Empower Your Team:** Use AI to free up time for your team to focus on more strategic, high-value activities that require a human touch.

❖ **Human Connection is Your Competitive Edge:** In a world of automation, it's genuine human connections that will set your business apart and build long-term loyalty.

❖ **Test, Adapt, and Stay Authentic:** Regularly test your AI systems and make adjustments to ensure they enhance, not erode, the customer experience.

MINDSHIFT EXERCISE #4:
The Human-AI Balance Audit

Spend 15 minutes auditing an AI-driven process (such as a chatbot or automated system) within your business. Identify the balance between AI and human interaction:

- ❖ Where does AI effectively handle repetitive tasks?
- ❖ Where does human empathy and judgment enhance the process?
- ❖ Find one opportunity to enhance AI's role and one where human interaction is needed more.

Action Step: Create a "Balance Map" for your organization and adjust processes to optimize the blend of AI efficiency and human touch.

Striking the perfect balance between AI efficiency and human empathy is essential, but how do you seamlessly implement AI across your entire business? In the next chapter, you'll discover how to integrate AI into your organization step by step, without overwhelming your team or losing the human touch that sets your business apart.

Section Two:
Implementing AI Across Business Functions

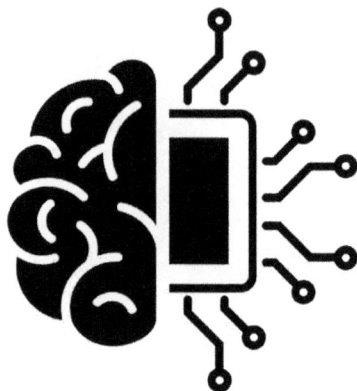

Here's where AI gets real. You've seen the potential—now it's time to bring it into every corner of your business. In this section, we'll break down how to seamlessly weave AI into your organization's DNA, from leadership to sales to customer service. It's not about throwing AI at every problem—it's about understanding where it adds the most value and how to use it to enhance, not replace, the human element in your business. By the end, you'll have a blueprint for scaling AI's impact across all areas, without losing what makes your business uniquely yours.

Chapter 5:

AI Mastery: The Four Stages of Seamless Integration

AI integration isn't a destination — it's an evolving journey that transforms your entire organization.

AI integration is no longer a question of "if" but "how." Leaders who understand how to leverage AI effectively across every department are the ones that will create scalable, future-proof organizations. And here's the kicker: it's not about adopting AI for the sake of technology. It's about solving real business challenges in a smarter, faster, and more sustainable way.

In this chapter, we're going to dive into the four levels of AI integration and explore how each one builds on the next, creating a seamless approach that drives performance and growth across your entire organization.

Level 1: Conversational AI (LLMs)

At the foundational level, we have **Conversational AI**, powered by large language models (LLMs) like ChatGPT, Claude, and Gemini. These tools are game-changers when it comes to interacting with customers and teams. They can handle tasks ranging from customer service inquiries to internal communication, providing fast, contextual responses that are light years beyond the old chatbot systems.

So, what does Level 1 AI integration look like in practice?

1. **Customer Service**: LLMs can handle initial inquiries, answer frequently asked questions, and provide 24/7 support. This not only reduces the burden on your customer service teams but also improves response times and customer satisfaction.

2. **Internal Support**: Whether it's HR, IT, or employee onboarding, LLMs can provide quick answers to routine questions, allowing your teams to focus on higher-value tasks.

The key here is that Conversational AI enhances speed and accessibility. It can handle the basic stuff so your human team can focus on more strategic work. But remember, this level should still be monitored, because while AI can manage routine questions, it's not yet capable of delivering empathy or handling complex situations.

Level 2: Industry-Specific AI Tools

The next level focuses on **Industry-Specific AI Tools**. These specialized solutions go beyond general-purpose AI by tackling the unique challenges of your specific sector—for example, whether that's healthcare, retail, manufacturing, or finance.

Here's how Level 2 AI integration plays out in different industries:

- **Marketing**: AI-powered tools can optimize content generation, provide predictive analytics, and offer precise audience segmentation. For example, AI can help you analyze customer data and predict future buying behavior, allowing your marketing team to create highly personalized campaigns.

- **Healthcare**: AI can assist with patient diagnostics, treatment plans, and predictive analytics, offering better patient outcomes and efficiency while navigating complex data and regulations.

- **Finance:** In the finance sector, AI tools can automate fraud detection and improve investment strategies by analyzing massive datasets and spotting patterns that might otherwise go unnoticed.

This level of AI integration allows you to go deeper into the specific needs of your industry, making smarter, data-driven decisions. It's about enabling your team to accomplish things that are too complex or time-consuming for humans alone.

Level 3: AI and Team Integration

At **Level 3**, AI moves beyond automation—it becomes a strategic partner. This is where you integrate AI deeply into your teams, enabling them to work faster and smarter by collaborating with AI.

Here's what that looks like in action:

- **AI as a Co-Worker**: AI can act as a brainstorming partner, generating ideas or content drafts that human teams refine. For instance, in sales, AI can prioritize leads based on predictive analysis, allowing your team to focus on closing deals.

- **Data-Driven Decision Making**: AI can identify trends, spot patterns, and provide real-time recommendations. Your team can then use this data to make informed decisions that are based on more than just gut feelings.

The key here is collaboration. When your team sees AI as an ally rather than a competitor, productivity skyrockets. AI takes over repetitive tasks, freeing your team to focus on strategic, high-value work.

Level 4: Workflow Automation with Zapier.com and Make.com

At **Level 4**, we enter the realm of **Workflow Automation**. This is where AI doesn't just handle individual tasks but automates entire processes, streamlining operations across your organization. Tools like **Zapier.com** and **Make.com** allow you to connect your CRM, email marketing, invoicing, and project management tools, automating workflows without the need for custom code. At this level, AI is integrated directly into your business processes, making operations run smoother with less human oversight.

Here's how Level 4 plays out in real-world scenarios:

- **Connecting Your Tools**: Zapier and Make can integrate your CRM, email marketing, invoicing, and project management tools. For example, you can set up an automation that sends a personalized email whenever a new customer signs up, adds them to your CRM, and triggers a task for your sales team to follow up.

- **Streamlining Repetitive Tasks**: Say goodbye to manual data entry and repetitive processes. These tools automate the grunt work, from generating reports to updating customer records, freeing up your team to focus on more strategic efforts.

- **Scaling Operations**: As your business grows, automation allows you to scale operations without dramatically increasing your workforce. With AI automating workflows, your team can handle more tasks with less manual effort, improving productivity and efficiency.

Level 4 is where you bring it all together—your Conversational AI, industry-specific tools, and AI-augmented teams are supported by an infrastructure that automates and connects every part of your business.

At this point, AI isn't just a tool your team uses—it's embedded into the very fabric of how your business operates.

Quick Win:
AI Integration Assessment

Choose one department in your business and conduct a quick AI integration assessment.

List out the current tasks, identifying which are repetitive or data-heavy.

Then, brainstorm how each level of AI integration (Conversational AI, Industry-Specific Tools, Team Integration, and Workflow Automation) could potentially improve these tasks.

This exercise will give you a concrete starting point for implementing AI in a strategic, layered approach.

Key Takeaways:

❖ **Conversational AI (Level 1):** Use tools like ChatGPT and Gemini to manage routine tasks and free up your team for more strategic work.

❖ **Industry-Specific AI Tools (Level 2):** Deploy AI solutions tailored to your sector to improve efficiency and data-driven decision-making.

❖ **AI and Team Integration (Level 3):** AI is a collaborator, not a replacement—integrate it with your team to enhance decision-making and productivity.

❖ **Workflow Automation (Level 4):** Automate repetitive processes to scale operations, improve efficiency, and reduce manual work.

MINDSHIFT EXERCISE #5:
The AI Integration Map

Select a department in your organization and categorize its tasks into four AI integration levels:

- ❖ **Conversational AI:** What simple tasks could AI handle (e.g., basic customer inquiries)?

- ❖ **Industry-Specific Tools:** Where could specialized AI tools be most beneficial (e.g., sales analytics, predictive maintenance)?

- ❖ **AI-Team Collaboration:** How could AI work with your team to enhance productivity?

- ❖ **Workflow Automation:** Which repetitive processes could be fully automated?

Action Step: Develop an integration plan by selecting one task from each category and implementing AI solutions in those areas. With a clear understanding of AI integration across different levels of your organization, it's time to explore how AI can elevate your leadership and decision-making. Get ready to unlock smarter, faster decisions that align AI's power with your strategic vision in the next chapter.

Chapter 6:

AI for Leadership and Decision-Making

AI doesn't replace leadership — it elevates it. The best leaders use AI to enhance their vision, not outsource it.

Every leader faces difficult decisions where the stakes are high. Wouldn't it be incredible if you could bring in a tool that processes huge datasets, identifies patterns, and provides insights faster than any human could? That's where AI comes in—but it's important to use it the right way. AI is here to assist, not replace leadership judgment.

This chapter will guide you through how to strategically integrate AI into your decision-making processes, enhancing your leadership capabilities without sacrificing the human element.

AI: A Game-Changer for Leadership Decisions

One of AI's most significant strengths is its ability to analyze large datasets in real time, distilling them into actionable insights. This isn't just a technological upgrade—it's a leadership transformation.

How AI Supports Leadership:

- **Data-Driven Insights:** AI pulls from sources like internal reports, customer feedback, and market trends to provide a clearer picture of your business environment. It helps you spot trends, predict customer needs, and understand what's truly going on in your operations.

- **Risk Management:** AI can identify risks that may not be immediately obvious. From financial risks to operational inefficiencies, AI-driven insights allow leaders to take proactive action before minor issues turn into major problems.

- **Operational Efficiency:** AI pinpoints inefficiencies, such as bottlenecks in your supply chain or staffing imbalances, enabling faster adjustments that save time and money.

But here's the key: AI doesn't replace your judgment. It's the flashlight that shines light on hidden insights, but you're still the one holding the flashlight, making the call on how to act.

Maintaining the Human Element in AI Decision-Making

While AI can process data quickly and offer insights, it can't bring empathy or human intuition into play. Leadership requires more than just interpreting numbers—it's about understanding the people behind those numbers. Empathy, ethics, and intuition are areas where AI can't compete.

Here's how to ensure you maintain the human touch while leveraging AI:

- **Context Matters:** AI may suggest cost-cutting by reducing staff in a department, but as a leader, it's up to you to weigh the human impact of that decision—on team morale, long-term productivity, and company culture.

- **Ethical Considerations:** AI doesn't have a moral compass. Decisions like hiring, firing, or customer policy changes require ethical judgment, and leaders need to ensure that AI-driven insights are ethically sound.

- **Listen to Your Team:** AI can identify patterns, but it can't replicate the nuanced understanding that your team brings. Incorporating feedback from your team leads to more balanced and informed decisions.

Implementing AI at Different Leadership Levels

AI isn't just for executives—leaders at all levels can leverage AI to make more informed, data-driven decisions.

- **Executive Leadership:** Use AI to gain insights into big-picture strategy, forecasting, and long-term planning. AI can analyze market trends and customer data, helping executives make decisions with a full view of potential risks and rewards.

- **Mid-Level Management:** AI tools help mid-level managers optimize operations by spotting inefficiencies, tracking performance, and suggesting improvements. For instance, AI could identify a drop in team productivity or suggest better resource allocation.

- **Team Leaders:** AI can assist with project management by flagging potential delays or shortages, allowing team leaders to make adjustments before problems arise.

Across all levels, AI enhances decision-making without replacing the essential human judgment that leaders bring to the table.

AI-Enhanced Leadership Decision-Making Framework

How to combine AI insights with leadership intuition:

1. **Define the Problem or Opportunity:** What decision are you facing?

    Example: Expanding into a new market.

2. **Gather AI Insights:** What does AI say about the problem? *Example*: AI shows customer demand is growing, but market saturation is a risk.

3. **Consider the Human Element:** What are the ethical or emotional impacts of this decision? *Example*: Cultural fit and customer loyalty will play a huge role in this expansion.

4. **Make the Decision:** How will you balance AI insights with human judgment?

    Example: AI supports expansion, but human intuition suggests focusing on a niche market segment.

5. **Monitor and Adjust:** Continuously monitor the results and gather feedback from both AI and human sources.

Quick Win:
Start an AI-Assisted
Decision Journal

For your next major decision, document the problem, the AI insights gathered, and the final decision made.

Include notes on how you balanced AI recommendations with your own judgment.

After a month, review the outcomes. This will help you refine your decision-making process.

Key Takeaways:

❖ **AI is a Tool, Not a Replacement:** Use AI to uncover trends, manage risks, and improve efficiency—but remember, it's your job to interpret and act on the insights.

❖ **Empathy is Still Key:** AI can't replicate empathy or ethical decision-making. Leadership requires a balance of data-driven insights and human intuition.

❖ **Implement AI at All Leadership Levels:** AI isn't just for the C-Suite—leaders at all levels can use AI to make smarter, faster decisions.

❖ **Monitor and Iterate:** Continuously review and refine your decisions, using both AI data and human feedback.

MINDSHIFT EXERCISE #6:
The Decision Dashboard

Review a recent leadership decision and ask yourself:

- ❖ What additional insights would AI have offered for this decision?

- ❖ How did your final decision align with or differ from AI's potential insights?

- ❖ What human elements (intuition, ethics, empathy) influenced your decision beyond data?

Action Step: Build a "Decision Dashboard" that integrates AI data into your leadership process, helping you combine AI-driven insights with human judgment for future decisions.

You now know how AI can supercharge your leadership decisions, but AI isn't just about data—it's also about emotional intelligence. Up next, we'll explore how AI can enhance human interactions, boosting both productivity and empathy to build stronger relationships with your team and clients.

Chapter 7:

AI and Emotional Intelligence (EQ) – A Winning Combination

The real power of AI isn't in its algorithms — it's in its ability to enhance our emotional intelligence and deepen human connections.

When it comes to business, we know that success is about more than just crunching numbers or implementing efficient processes. The most successful leaders and companies understand the power of **emotional intelligence (EQ)**—the ability to connect with people on a human level, show empathy, and build trust. But what happens when you add AI to the mix? Can AI complement emotional intelligence? Can you integrate EQ into AI-driven interactions to create better outcomes for your team and clients?

The answer is yes—but only if you approach it the right way. AI can enhance certain aspects of your business, but it will never replace the human side of things. When you combine the analytical power of AI with the emotional insights of EQ, that's when you create real magic in your business. Let's explore how this winning combination works and how you can use it to build stronger relationships with both your team and your clients.

How AI Enhances Emotional Intelligence

At first glance, AI and EQ might seem like oil and water. AI is all about data, algorithms, and automation, while EQ focuses on feelings, reading non-verbal cues, and building relationships. But here's the twist: when used together, AI can actually enhance EQ. It can handle data at lightning speed, leaving you free to focus on the human element.

AI Analyzes Data; Humans Analyze Emotions

AI is fantastic at processing mountains of data—tracking customer behavior, analyzing engagement metrics, and running surveys. It can tell you what's happening, but it can't explain why. That's where EQ steps in. Emotional intelligence allows you to interpret the "why" behind your customer's actions or your team's dynamics. AI gives you the facts; EQ gives you the insights.

Predictive Analytics Meets Emotional Insight

AI can predict trends and forecast customer behaviors, but it won't tell you how someone feels in the moment. Your ability to pick up on emotional cues—whether verbal or non-verbal—lets you respond in a way that feels personal and genuine. When you combine AI's predictive power with your emotional insights, you get the best of both worlds. Now, you're not just predicting behaviors; you're anticipating needs.

AI Provides Context; EQ Navigates It

Picture this: AI tells you a customer has been browsing your website for a long time without buying anything. That's useful data, but it doesn't tell you why. EQ fills in the gaps. When your customer service team reaches out with empathy—acknowledging potential frustrations and offering solutions—you turn AI's data into meaningful human interactions. AI shows you where to look; EQ helps you act on it.

Real-World Example: AI and EQ in Action

Let's say your customer service team is using AI to handle FAQs and simple requests. Sounds efficient, right? But what happens when a frustrated customer gets stuck in a loop with the chatbot, feeling like they're not being heard?

Here's how you fix that:

- **AI's Role:** AI continues to handle routine inquiries, but as soon as it detects emotional words like "angry" or "upset," it triggers a human agent to step in.

- **Humanizing the Response**: Even AI responses should sound human. Instead of a cold, "Your request is being processed," try, "I see this is frustrating. We're here to help, and a team member will reach out shortly."

This hybrid approach uses AI to boost efficiency without losing the emotional connection that builds trust.

How to Keep AI and EQ Balanced

1. Use AI for Routine Tasks

Let AI handle the repetitive stuff: FAQs, follow-up emails, and basic data analysis. This frees up your team to focus on the interactions where empathy and creativity are needed.

2. Humanize Automated Responses

Automated doesn't have to mean robotic. Make sure your AI-generated communications reflect empathy. "Thanks for reaching out! We know this can be frustrating, and we're on it," sounds a whole lot better than a cold, "Your request has been received."

3. Transition to a Human When Necessary

Know when to pass the baton. If a customer is repeatedly asking for help or using emotional language, AI should trigger a handoff to a real person. That's how you keep things personal when it matters most.

4. Review and Improve AI Interactions Regularly

Regularly audit your AI systems for moments where emotional intelligence is lacking. Are customers getting frustrated? Are there areas where a human touch could improve the experience? Gather feedback and tweak your systems to ensure that AI never sacrifices empathy for efficiency.

Quick Win:
EQ-AI Interaction Audit

Pick a recent AI-driven customer interaction (like a chatbot conversation or automated email response).

Review it through an EQ lens. Identify moments where the AI could have shown more empathy or where a human touch would have made a difference.

Then brainstorm three ways to infuse more EQ into the interaction. This exercise helps you find the balance between AI efficiency and human emotional intelligence.

Key Takeaways:

❖ **AI Supports EQ, But Doesn't Replace It:** AI provides data and speed, but it's up to your team to layer in the emotional intelligence that builds trust and relationships.

❖ **Humanize Your AI Communications:** Ensure your automated responses feel personal and aligned with your brand's voice.

❖ **Balance AI and Human Interaction:** Use AI for routine tasks, but know when to step in with human empathy and understanding. This balance keeps your business efficient without losing the personal connection that makes all the difference.

MINDSHIFT EXERCISE #7:
The AI-EQ Synergy Audit

Audit an AI tool you use in customer service or employee interactions:

❖ Where does AI help solve issues quickly but lack emotional intelligence?

❖ How can you better integrate human empathy into AI-driven interactions?

❖ Identify one customer or team scenario where AI should "hand off" to a human for a more emotionally intelligent resolution.

Action Step: Implement "AI-EQ Synergy" checkpoints in your business, ensuring AI handles routine tasks while humans take over emotionally sensitive or complex situations.

Combining AI with emotional intelligence is a powerful formula for success. But to truly capitalize on AI's potential, your organization needs a solid adoption strategy. In the next chapter, we'll cover how to get your entire team on board with AI, ensuring it becomes an integrated part of your company culture.

Chapter 8:

Organizational AI Adoption Strategies

Successful AI adoption isn't about upgrading your technology; it's about upgrading your entire company culture.

AI isn't just another IT project—it's a game-changer that's reshaping everything. But here's the kicker: to truly harness AI's power, your entire organization needs to be on board, not just the tech team. This chapter is your roadmap to making AI a seamless, powerful driver of innovation and performance across your company.

AI Mindshift: Getting Everyone on Board

Let's cut to the chase: AI isn't just some IT department's pet project anymore. It's the future of business, and if your whole company isn't on board, you're already falling behind. But here's the thing — you can't just drop a bunch of AI tools into people's laps and expect magic to happen. You've got to build a culture where AI is as natural as checking your email.

Why AI Literacy Matters

Think about it. Would you hand someone the keys to a Ferrari if they've never driven before? Of course not. Same goes for AI. Your team needs to understand this stuff, not just the tech whizzes. When everyone gets it, that's when AI becomes your secret weapon.

Speak Their Language

Forget the tech jargon. Show each department how AI makes their job easier, faster, or just plain better.

- Marketing folks? Show them how AI can predict which campaigns will actually work.

- Sales team? Let them see how AI can spot the hottest leads.

- Customer service? Demonstrate how AI can handle the boring stuff so they can focus on what matters.

Provide Role-Specific Training

One-size-fits-all training is a waste of everyone's time. Give each team the AI training they need:

- Finance needs to know about fraud detection and forecasting.

- HR should learn about smarter hiring and keeping employees happy.

- Product development? AI for market trends and design optimization.

- Operations: Showcase AI's impact on supply chain optimization.

- Shameless (Easter egg) Promo: Contact me about customized presentations designed for your specific needs and outcomes. Watch my speaker reel at ProfitRichResults.com

Make It Fun, Not Scary

AI should be a toy to play with, not a monster under the bed. Create a sandbox where people can mess around with AI tools without fear. Reward the creative ideas. Hell, throw an "AI Olympics" if you have to. Make it exciting.

Find Your AI Champions

In every department, there's someone who gets excited about new tech. Find them, nurture them, and let them loose. These are your AI evangelists. They'll do half the work for you, spreading the AI gospel better than any top-down mandate ever could.

Show Off the Wins

Nothing sells like success. When AI helps someone crush their targets or solve a nagging problem, shout it from the rooftops. Share these stories everywhere – company newsletters, meetings, even in the break room. Make AI the hero.

Address the Elephant in the Room

People are going to worry about AI taking their jobs. Don't dodge it. Talk about it head-on. Show how AI is a tool to make them better at their jobs, not replace them. Be honest about changes but focus on the opportunities.

The Payoff

When you nail this AI literacy thing, magic happens:

- Silos crumble as teams start speaking the same AI language.

- Innovation explodes because everyone's thinking about how AI can push boundaries.

- Decisions get sharper because people know how to use AI insights.

- Your company becomes future-proof, ready to jump on whatever tech comes next.

The Bottom Line

Building AI literacy isn't a one-and-done deal. It's a constant effort. But here's the kicker – it's worth every ounce of sweat. You're not just teaching people about some new tech tool. You're reshaping how your entire company thinks and works.

Get this right, and you won't just be riding the AI wave – you'll be making the waves that others try to catch.

Remember, the goal isn't to turn your marketing manager into a data scientist. It's about making everyone comfortable enough with AI that they're excited to use it, not scared of it. Do that, and watch your company transform before your eyes.

Now go make it happen. The AI revolution waits for no one.

Quick Win:
AI Champion Kickstarter

Here's a fun one:

Host an "**AI Champion Kickstarter**" lunch. Invite a mix of tech-savvy folks and those who might be more AI-hesitant from different departments.

Over sandwiches (or pizza, I'm not picky), have everyone share one way AI could make their job easier.

You'll be amazed at the ideas that pop up. Plus, you'll spot potential AI champions who can help drive adoption across your organization. It's a win-win: free lunch and a jumpstart on your AI strategy. Who said innovation can't be delicious?

Key Takeaways:

❖ **Build AI literacy across all departments** with role-specific training to ensure employees understand how AI can enhance their daily work.

❖ **Empower AI champions** to lead the way by providing advanced training and encouraging them to share their knowledge across teams.

❖ **Foster cross-functional collaboration** by breaking down silos and encouraging different departments to come together and innovate using AI.

❖ **Engage the entire organization** by having leadership communicate the importance of AI, sharing success stories, and building a feedback loop for continuous improvement.

MINDSHIFT EXERCISE #8:
AI Adoption Playbook

Take a closer look at how AI adoption is unfolding across your organization. Answer these questions:

- ❖ Which departments are most receptive to AI, and why?
- ❖ Where are the biggest barriers to AI adoption, and what is causing resistance?
- ❖ What role-specific training or tools could help each team become more comfortable with AI?

Action Step: Develop a tailored "AI Adoption Playbook" that includes role-specific training plans and success stories from early adopters in your company.

Now that you have a roadmap for getting your team on board, it's crucial to be aware of the common pitfalls that can derail your AI initiatives. In the next chapter, we'll uncover the most common AI mistakes businesses make—and, more importantly, how you can avoid them to stay on track.

Section Three:
Navigating Pitfalls and Avoiding Common AI Challenges

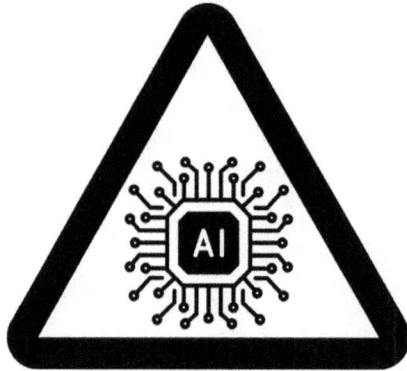

AI can open up incredible opportunities, but it's not without its landmines. This section is about ensuring you don't step into the pitfalls that can derail your efforts before they even get off the ground. Whether it's ethical risks, over-reliance on automation, or failing to align AI with your business goals, we'll tackle the common mistakes and show you how to avoid them. Because when it comes to AI, a smart strategy isn't just about what you do—it's about what you don't do.

Chapter 9:

Common AI Pitfalls and How to Avoid Them

AI can take you far, but without clear goals and human oversight, it can take you in the wrong direction faster than ever.

Let's start with a reality check: Whether you've officially adopted AI tools in your business or not, I can guarantee that some of your staff are already using them. They're experimenting with ChatGPT to draft emails, running social media campaigns with AI-generated content, or maybe even using AI tools for data analysis—all without you even knowing.

And here's the kicker: without best-practice guidelines or acceptable use policies in place, you're leaving your business exposed.

AI is a powerful tool, but it comes with challenges. The biggest risk in adopting AI isn't in making technical mistakes—it's in failing to align AI initiatives with your broader business goals. Many businesses jump into AI without understanding its limitations or the potential pitfalls.

The Danger of Over-Reliance on Automation

One of the most common mistakes businesses make is relying too heavily on AI for tasks that require human nuance. AI is excellent at automating repetitive tasks and analyzing large datasets, but it lacks the emotional intelligence and empathy that humans bring to the table.

For instance, an e-commerce company adopted an AI-powered chatbot to handle customer service. While it excelled at answering FAQs, it struggled with more complex, emotionally charged inquiries. Customers dealing with sensitive issues, such as delayed deliveries or payment problems, felt frustrated by the chatbot's impersonal responses, leading to negative reviews and reputational damage.

Or consider my own experience. I've always wanted to implement a chatbot for my speaking and agency websites, but the idea of manually scripting responses felt daunting. The early versions weren't cutting it—they just didn't engage visitors meaningfully. That's when I set up a smarter AI chatbot through Zapier.com, using documents about my business and services.

Now, when the AI detects words like "frustrated" or "confused," it immediately triggers a human intervention. The chatbot works great for simple inquiries, but when it senses the need for human empathy, it hands things off to my team. This balance keeps the AI efficient without losing the personal connection that customers need.

These real-world examples highlight a crucial point: AI can be incredibly powerful when used correctly, but without proper oversight and human involvement, it can go from helpful to harmful fast.

Lack of Clear Objectives for AI Implementation

Jumping into AI adoption without a clear purpose is another major pitfall. Too many organizations rush to implement AI tools without first identifying the problems they are solving or how AI fits into their long-term business strategy. This results in wasted resources and frustration.

A mid-sized retail chain implemented AI-driven inventory management without first clarifying its goals. While the AI improved stock management in some locations, it led to stockouts in others because it wasn't tailored to regional demand. The lack of a clear objective and customization resulted in supply chain inefficiencies and customer dissatisfaction.

Before adopting any AI tool, it's critical to start with the problem you're trying to solve. Align your AI strategy with measurable business outcomes, such as reducing costs, improving customer engagement, or enhancing decision-making. By setting clear objectives, you can ensure that AI contributes positively to your bottom line.

Poor Data Management and Quality

Data is the foundation of any successful AI initiative, but poor data management can quickly undermine your efforts. If your data is incomplete, biased, or outdated, your AI systems will struggle to deliver accurate or valuable insights.

A financial services firm implemented an AI-driven customer segmentation model based on outdated demographic data. The AI generated misleading profiles, targeting campaigns at customers who no longer fit the company's buyer persona. This misalignment led to wasted marketing efforts and a decline in campaign effectiveness.

To avoid these issues, invest in data hygiene. Regularly clean, update, and verify your datasets to ensure your AI systems are working with reliable information. Establish a data stewardship program to maintain high data quality, ensuring your AI produces meaningful results.

AI Without Human Oversight

While AI is capable of processing vast amounts of data and generating insights, it still lacks the human judgment required for critical decisions. Businesses that rely too heavily on AI without human oversight risk making costly mistakes.

For example, an insurance company used AI to assess claims and automatically approve or deny payouts. Over time, the AI began rejecting legitimate claims due to an overly strict algorithm, leading to customer backlash and reputational damage. Without human oversight to review these decisions, the company faced legal challenges and lost customer trust.

It's crucial to maintain human involvement in high-stakes decisions. AI can provide recommendations, but human review is necessary for decisions that affect customer satisfaction, legal matters, or sensitive business processes. By combining AI with human oversight, businesses can mitigate risks while still benefiting from AI's efficiency.

Misalignment with User Experience

AI tools can dramatically impact user experience—both positively and negatively. Poorly designed AI interfaces can frustrate customers and damage brand perception, especially when they fail to meet user expectations.

A tech company implemented an AI-powered recommendation engine on its e-commerce platform. However, the recommendations were irrelevant to customers' browsing behavior, leaving users feeling disconnected and confused. As a result, the company experienced a drop in sales and customer engagement.

To prevent such failures, it's essential to test AI-driven tools with real users before launching them broadly. Gathering feedback and continuously refining AI outputs will help ensure that your AI aligns with customer needs and enhances their experience, rather than detracting from it.

Quick Win:
AI Pitfall Audit

Take a moment to audit your current AI tools by asking the following questions:

- ❖ Are we over-relying on AI for tasks that need human judgment?
- ❖ Do we have clear objectives for every AI tool in place?
- ❖ How clean and up-to-date is our data?
- ❖ Is there adequate human oversight in AI decision-making?
- ❖ Is our AI enhancing or detracting from the customer experience?

Identifying and addressing weak points in your AI strategy now will help you avoid future pitfalls and ensure that your AI initiatives are successful.

Key Takeaways

❖ **Avoid Over-Reliance on AI**: While AI can automate repetitive tasks, human oversight is crucial for decisions requiring empathy and complex judgment.

❖ **Align AI with Clear Business Goals**: Start with specific, measurable objectives to ensure your AI delivers value.

❖ **Maintain High-Quality Data**: Regularly clean and update datasets to ensure that AI systems deliver accurate insights.

❖ **Ensure Human Oversight**: Human involvement is essential for high-stakes decisions, even when AI is providing recommendations.

❖ **Focus on User Experience**: Test AI tools with real users to ensure they enhance—not hinder—the customer journey.

MINDSHIFT EXERCISE #9:
AI Pitfall Prevention Plan

Reflect on potential pitfalls in your AI strategy. Ask yourself:

❖ Are you over-relying on AI in any areas where human judgment is still essential?

❖ Are your AI tools aligned with clear business goals and metrics?

❖ How robust is your data quality? Are there gaps, biases, or outdated information?

Action Step: Create an "AI Pitfall Prevention Plan" by identifying the top three risks in your current AI processes and drafting steps to mitigate each one.

Knowing the pitfalls is half the battle, but how do you ensure your AI initiatives are ethical, unbiased, and fair? In the next chapter, we'll explore the importance of creating ethical AI practices that protect your business and your customers from legal and reputational risks.

Chapter 10:

Navigating Data Privacy, Bias, and Fairness — Creating Ethical AI Practices

Ethical AI isn't a roadblock to innovation — it's the highway to sustainable, trust-driven growth.

Disclaimer: Before we dive in, let me make one thing crystal clear: I'm not a lawyer, and this chapter (and entire book) isn't legal advice. I'm bringing up these topics because they're critical to your success in using AI responsibly. When it comes to legal compliance, consult your legal team or an expert in data protection laws. Trust me, you'll thank me later.

AI Gone Wrong: Real-Life Horror Stories

Let's start with a few cautionary tales that highlight why data privacy, bias, and fairness are so critical. These aren't made-up stories—they're real-world examples of AI gone completely sideways, making headlines and causing massive headaches.

- "McDonald's Drive-Thru AI Experiment Crashes and Burns"

- "NYC Chatbot Encourages Business Owners to Break the Law"

- "Air Canada Loses Court Battle After Chatbot Lies About Bereavement Policy"

- "Amazon's AI Recruitment Tool Scrapped for Discriminating Against Women"

- "IBM's Watson for Oncology Canceled After $62 Million and Unsafe Cancer Treatment Advice"

- "Facial Recognition Fail: Amazon's AI Mistakes 28 Congresspeople for Criminals"

- "Zillow's House-Flipping AI Disaster Leads to $500 Million Loss and 2,000 Layoffs"

These stories show how easily AI can spiral out of control if it's not managed correctly. We're talking about financial loss, legal consequences, and PR nightmares—none of which any business wants to deal with. And these aren't just problems for the big players. If you're using AI in your business without thinking through the ethical implications, you're putting yourself at risk.

Transparency and Explainability in AI Decision-Making

One of the biggest complaints about AI is its "black box" nature. AI makes a decision, but no one knows how or why. This kind of opacity erodes trust, and that's a problem.

1. **Embrace Explainable AI (XAI)**: Wherever possible, use AI systems that explain their decisions in plain language. If an AI denies someone a loan, it should be able to break down the "why" behind it. Customers trust you more when they know what's going on.

2. **Own Your Decisions**: Even if an AI system is making decisions, you're still accountable. If something goes wrong, you can't pass the blame to the AI. Take ownership, fix the problem, and communicate clearly how you're going to prevent it from happening again.

3. **Be Open About AI Usage**: Don't hide behind your technology. Tell your customers and stakeholders how you're using AI, what data you're collecting, and how you're pro-

tecting their information. Transparency builds trust, and trust is your most valuable currency in the age of AI.

AI's Impact on Employment and the Workforce

AI is shaking up the workforce, and while it's a game-changer for productivity, it also brings concerns about job displacement.

1. **Upskilling, Not Replacing**: Instead of fearing AI as a job-killer, embrace it as a tool for empowering your employees. Invest in upskilling your team so they can work alongside AI, leveraging the tech to take on more strategic, creative roles.

2. **New Opportunities**: AI is also creating entirely new job categories. From AI oversight to ethical governance, there's a growing need for roles that didn't exist just a few years ago. Focus on building your team around these new opportunities.

3. **Be Transparent with Your Team**: Don't blindside your employees. Be upfront about how AI will affect their jobs and make them part of the journey. When your team understands the vision, they're more likely to buy into it.

Environmental Implications of AI Implementation

AI has its environmental footprint, especially when it comes to the energy needed to process massive amounts of data. But this can't be an afterthought.

1. Go Green with AI: Some companies are pioneering the "Green AI" movement, optimizing their AI systems to minimize energy consumption. Efficiency matters—not just for the bottom line, but for your brand's sustainability efforts.

2. Optimize for Efficiency: Simple tweaks in how you build and deploy AI can reduce its environmental impact. You can cut resource use without sacrificing performance, which is a win-win for your business and the planet.

AI and Social Responsibility

AI's impact isn't limited to business; it has ripple effects on society as a whole. The way you deploy AI has social consequences, and that's a responsibility you can't ignore.

1. **Ethical Deployment**: You need to think beyond profits. How is your AI affecting your customers, your community, and society? AI that perpetuates inequality or social harm will eventually hurt your brand. Take the long view.

2. **AI for Good**: AI can be a powerful force for positive change. Think about using AI to solve bigger issues like healthcare, education, or environmental sustainability. Doing good is good for business, too.

Ethical Considerations in AI Research and Development

How you develop AI matters just as much as how you deploy it.

1. **Diversity in R&D**: AI is only as good as the people building it. If your development teams all look the same, your AI will reflect that. Diverse teams bring diverse perspectives, which leads to more balanced and ethical AI solutions.

2. **Ethical Testing**: Before you roll out any AI solution, it's crucial to test for ethical issues. You don't want to find out your AI is biased or unsafe after it's too late. Run simulations, check for biases, and have a plan for constant improvement.

A Practical Framework for Ethical AI Implementation

To help you keep your AI systems on the ethical side of the fence, here's a practical framework you can use in your business:

1. **Audit Your AI Regularly**: Regularly review your AI systems for bias, data privacy issues, and transparency gaps.

2. **Create an Ethical AI Task Force**: Set up a dedicated team responsible for keeping your AI ethical.

3. **Involve Your Stakeholders**: Employees, customers, and communities should have a voice in how AI is implemented.

4. **Test for Fairness**: Simulate different outcomes to ensure your AI isn't favoring or harming any group.

5. **Monitor and Adapt**: This isn't a "set it and forget it" situation. Your AI needs continuous monitoring and tweaks to stay ethical.

Global Perspectives on AI Ethics

Ethical AI isn't a one-size-fits-all proposition. Different cultures and regions have their own approaches, and if you're working on a global scale, you need to understand them.

- **Europe**: The EU's regulations, like GDPR, are some of the most stringent when it comes to AI and data privacy. If you're doing business in Europe, compliance is key.

- **United States**: The U.S. has a more flexible regulatory environment, but that doesn't mean you can be lax. Self-regulation and ethical AI frameworks are gaining traction.

- **Asia**: In China, AI is often used for public surveillance, which raises different ethical questions compared to Western nations. It's a rapidly advancing region in AI, but the ethics vary widely.

Future Challenges in Ethical AI

As AI becomes more sophisticated, new ethical challenges will arise. Here are a few that are already on the horizon:

- **Autonomous Decision-Making**: What happens when AI starts making decisions without human oversight? That's a future we need to prepare for.

- **Deepfakes**: AI-generated fake media is becoming a huge issue, from misinformation to security threats.

- **AI in Warfare**: The potential for AI weaponization raises a whole host of ethical and legal concerns that we haven't fully grappled with yet.

Regulatory Landscape: Present and Future

Regulations are playing catch-up with AI. Here's where things stand:

1. **Current Regulations**: Laws like GDPR and CCPA are just the beginning. More comprehensive AI regulations are on the way, and staying ahead of them will save you headaches down the road.

2. **Be Proactive**: Don't wait for the government to force your hand. Implement ethical AI practices now, so when regulations do arrive, you're already prepared.

Stakeholder Engagement in Ethical AI

Ethical AI impacts everyone—employees, customers, communities—and they all deserve a seat at the table.

1. **Internal Buy-in**: Your employees need to understand how AI affects them and be part of the decision-making process.

2. **Customer Transparency**: Be clear with your customers about how you're using AI, especially when it comes to their data and privacy.

3. **Social Responsibility**: Think about how your AI choices affect the broader community, including the potential environmental and social impacts.

Ethical AI Governance Structures

Building an ethical AI governance structure within your organization is non-negotiable.

1. **Ethics Committees**: Set up an internal committee to oversee AI ethics and keep things on track.

2. **Code of Conduct**: Create a clear code of conduct for AI use in your organization, and make sure everyone understands it.

3. **Continuous Education**: AI ethics evolve, so your team's understanding should too. Regular training is a must.

Ethical AI Auditing

You can't manage what you don't measure, and that applies to AI ethics.

1. **Internal Audits**: Regularly review your AI systems to ensure compliance with ethical standards.

2. **External Audits**: Bring in third-party experts to provide unbiased feedback and certify that your AI is operating fairly and transparently.

Quick Win:
Ethical AI Escape Room

Here's a fun, practical exercise for your team. Set up an **"Ethical AI Escape Room"** challenge, where you create a hypothetical scenario involving AI making biased decisions or violating data privacy.

Give your team 30 minutes to "escape" by identifying the ethical issues and proposing solutions. It's like a mini-hackathon with a purpose, and it makes ethics training more engaging and memorable.

Key Takeaways:

❖ **Ensure Data Privacy and Compliance**: Understand the regulations, anonymize data, and use enterprise plans with increased security to safeguard your data.

❖ **Identify and Mitigate Bias**: Regularly audit your AI systems for bias and ensure that high-stakes decisions involve human oversight.

❖ **Prioritize Transparency**: Implement explainable AI and clearly communicate with customers and stakeholders about how AI is being used.

MINDSHIFT EXERCISE #10: AI Ethics Audit

Conduct an audit of your AI tools and processes, focusing on ethics:

- ❖ Are there areas where your AI might introduce bias or unfairness?
- ❖ How are you ensuring transparency and explainability in AI-driven decisions?
- ❖ What steps can you take to improve data privacy and security?

Action Step: Implement an "AI Ethics Audit" framework to periodically check your systems for bias, data privacy concerns, and ethical transparency, and make adjustments as necessary.

With your ethical framework in place, the next step is preparing for when things go wrong. Crisis management is where AI truly shines. In the next chapter, we'll examine how AI can help you respond quickly and effectively in high-stakes situations where speed matters most.

Chapter 11:

AI in Crisis Management – When Speed Matters

In a crisis, AI isn't just a tool — it's a force multiplier that turns challenges into opportunities for leadership.

When a crisis hits, every second counts. Whether you're dealing with a PR nightmare, a security breach, or a supply chain disruption, the window for making the right decisions is small—and shrinking by the minute. In these moments, AI becomes your fast-acting, data-driven partner, helping you cut through the chaos and make critical decisions with clarity and precision.

But let's get one thing straight: AI isn't here to replace humans in a crisis. It's here to empower you to make better, faster decisions when it matters most. By analyzing vast amounts of data in real time and delivering actionable insights, AI gives you the edge in managing crises effectively.

Why Speed Matters in Crisis Management

In a crisis, there's no room for indecision. Every moment spent debating your next move could cost you millions, tarnish your reputation, or even jeopardize the survival of your business. AI eliminates the lag that often comes with human decision-making by providing instant access to critical data, allowing you to act quickly with confidence.

When everything goes wrong, human decision-making can get messy. We second-guess, get emotional, and debate—taking time we don't have. AI helps cut through that. It offers real-time analysis, gives you the ability to simulate various scenarios, and lets you act based on facts, not assumptions.

AI in Action: A Cybersecurity Case Study

Imagine this scenario: It's 2:37 AM, and Midwest Financial, a regional bank serving over 2 million customers, is under siege from a sophisticated ransomware attack. In the past, this would have been a disaster in the making—the IT team scrambling to contain the damage, possibly too late. But Midwest Financial had implemented an AI-powered cybersecurity system, and here's what happened instead:

1. **Detection**: Within seconds, AI identified the anomalous behavior and flagged it as a potential attack.

2. **Analysis**: In under a minute, AI analyzed the threat, identified it as a new strain of ransomware, and predicted its likely path of infection.

3. **Containment**: By the two-minute mark, AI had isolated the affected systems, preventing the ransomware from spreading to critical servers.

4. **Response**: Within five minutes, the AI initiated countermeasures, updated firewall rules, and blocked vulnerable ports to minimize the threat.

5. **Human Intervention**: When the IT team arrived 20 minutes later, they found the situation largely contained. AI had already generated a detailed report, allowing the team to focus on eliminating the ransomware and patching vulnerabilities.

The result? While some systems were affected, customer data remained secure. AI didn't replace the IT team—it empowered them. It bought them time, and in cybersecurity, time is more valuable than gold.

AI in Crisis Communication: Managing the Message

When a crisis occurs, it's not just about solving the problem—it's about managing the message. The way you communicate with customers, stakeholders, and the public can either salvage your reputation or destroy it. This is where AI shines in monitoring real-time social sentiment, news reports, and customer inquiries.

Imagine being able to adjust your messaging on the fly based on immediate feedback from AI-driven tools. AI can help you gauge how the crisis is unfolding in the public eye and guide you in crafting communications that calm fears, address concerns, and protect your brand before rumors spiral out of control.

Building AI into Your Crisis Management Plan

Now here's the catch: AI is only helpful in a crisis if you've integrated it into your systems beforehand. You can't afford to be learning how to use AI in the middle of a disaster. It needs to be part of your crisis management toolkit from the start, and your team must be trained and ready to trust the system when things go wrong.

1. Get Your Systems in Place

Start by identifying the areas where AI can make the biggest impact in a crisis. Whether it's real-time data analysis, cybersecurity, or customer communication, integrating AI ahead of time ensures you're ready to act quickly when disaster strikes.

2. Train Your Team

AI is a tool, not a crutch. Your team needs to know how to work with it—not against it. They should understand when to trust AI's insights and when to step in with human judgment. Crisis response is a partnership between AI's speed and human intuition.

3. Balance Speed with Human Judgment

AI is all about speed, but human judgment is still critical in crises. While AI handles the data-heavy tasks and provides quick decisions, your human team brings the empathy, creativity, and context needed to handle complex problems that technology can't fully grasp.

Quick Win:
Run an AI Crisis Drill

Don't wait for a real crisis to test your AI systems. Hold an "**AI Crisis Drill**" with your team.

Set up a hypothetical scenario—whether it's a PR disaster, product failure, or cybersecurity breach—and use your AI tools to analyze data and simulate responses.

This drill will show your team how AI can speed up decision-making and highlight any gaps in your crisis management strategy.

Key Takeaways

❖ **AI Accelerates Crisis Response**: By analyzing real-time data and automating routine tasks, AI allows your team to make faster, more informed decisions when time is of the essence.

❖ **Simulate Scenarios for Strategic Planning**: Use AI to predict the outcomes of various crisis strategies, helping you choose the best course of action before the crisis escalates.

❖ **Pre-crisis Integration is Critical**: AI must be part of your crisis toolkit before disaster strikes—don't wait until you're in the thick of a crisis to figure out how to use it.

❖ **Balance AI with Human Oversight**: AI handles the speed and data; humans handle the strategy and decision-making. Together, they make a powerful team for managing crises.

MINDSHIFT EXERCISE #11:
AI Crisis Response Drill

Run a simulated crisis (e.g., data breach, PR issue, supply chain disruption) and assess your AI's role in managing it:

❖ How quickly did AI identify the issue?

❖ How did AI's analysis inform the team's response?

❖ Was there a balance between AI speed and human judgment in resolving the crisis?

Action Step: Develop an "AI Crisis Response Drill" that you run quarterly to ensure your team and AI systems are ready for high-pressure situations.

By preparing your AI systems and blending them with human insight, your organization can respond to crises faster, smarter, and more effectively turning potential disasters into opportunities to show off your leadership chops.

And don't forget, my latest AI training resources are always available at:
ProfitRichResults.com/ai-training if you want to dig deeper into fresh perspectives on how AI can help you manage business challenges!

Alright, crisis averted (at least for now). But before you kick back, we've got to tackle something that might not seem glamorous but is absolutely necessary: AI compliance. I get it— navigating regulations might sound as exciting as watching paint dry, but getting this right will save you from some serious headaches—and even more serious fines.

Grab that extra coffee and gear up, because we're diving into the world of AI regulations. I promise we'll keep it as painless as possible!

Chapter 12:

Navigating the AI Compliance Maze

Compliance isn't just a box to check — it's the foundation of trust and sustainable growth in the AI era.

AI has transformed industries, but compliance is what will determine which businesses thrive and which face disaster. In the race to innovate, failing to meet regulations is like building a rocket without a safety check—it may soar, but the crash will be catastrophic. This chapter will guide you through the complexities of AI compliance, not as a burden, but as a strategic advantage that can set your organization apart. And remember my disclaimer… this is for educational purposes… yada, yada, yada..

AI regulations, also known as the wild wild west, is a rapidly changing landscape—one where businesses are still figuring out how to navigate shifting laws. Imagine a company using AI to make hiring decisions, only to find out their algorithm is inadvertently biased. The penalties? Not just fines, but a loss of trust that may never be regained. The regulatory sheriff *is* coming, and businesses that aren't ready will face the consequences.

You've heard the acronyms: GDPR, CCPA. These aren't just bureaucratic red tape—they're the gatekeepers of your customers' trust. GDPR fines can reach up to 20 million euros or 4% of your annual turnover, whichever is higher. If that doesn't make you nervous, it should. But compliance isn't just about avoiding fines; it's about showing your customers that you take their privacy seriously, which can set you apart from competitors still ignoring the rules.

For instance, the EU's AI Act is set to impose strict requirements for high-risk AI systems—think AI in healthcare, recruitment, or financial services. Non-compliance won't just mean legal penalties; it'll erode customer confidence. Remember, AI might be cutting-edge, but if it isn't ethical, it's unusable.

Here's your AI Compliance Checklist, a must-have for every business leader integrating AI:

1. **Data Privacy**: Ensure you're collecting only the data you need. If users can't easily delete their data, you're one click away from a lawsuit.

2. **Transparency**: Can your AI explain its decisions? If not, regulators won't be the only ones asking questions—your customers will, too.

3. **Fairness**: Audit your algorithms for bias regularly. It's not just about compliance; it's about building AI systems that don't perpetuate inequality.

4. **Security**: Treat your AI systems like Fort Knox. A breach isn't just a tech failure; it's a trust failure.

5. **Accountability**: When AI goes rogue, who's responsible? Here's a hint: it's you, and regulators will be knocking.

AI regulations are evolving as fast as the technology itself. In the near future, we'll see:

1. **Stricter Data Privacy Laws**: As AI's hunger for data grows, expect tighter controls on what can be collected and how it's used.

2. **Bias Audits and Certifications**: Just like environmental certifications, expect third-party audits ensuring your AI doesn't favor one group over another.

3. **AI-Specific Laws**: From employment impacts to decision-making, new laws will emerge to handle the unique challenges AI presents.

4. **Global Standards**: The world's moving towards unified AI regulations. If you're an international player, you'll need to comply with laws across multiple borders—simultaneously.

At the heart of AI compliance is trust. In a world increasingly dominated by algorithms, people will crave human connections more than ever. Compliance isn't just about avoiding fines; it's about proving to your customers that your AI systems operate with integrity and ethics. Show them that your business isn't just compliant but committed to responsible AI.

While your competitors are playing catch-up, your organization can stand out by championing ethical AI. Think of compliance as your business's 'Certified Ethical AI' label. As trust becomes a rare commodity, those who get compliance right will rise above, becoming the go-to players in a crowded market.

I'm switching the Quick Wins up a bit, and just adding them here.

Here's how to make compliance less of a headache and more of a team-building exercise: set up a **"Compliance Scavenger Hunt"** in your organization.

Challenge your team to identify where key AI regulations apply in your business operations. First to spot all the key regulations wins a prize—think "Skip the Next Meeting" card. Not only will this activity make compliance less dull, but it'll also help uncover any gaps you might have missed. Everyone wins.

Key Takeaways

❖ **Stay ahead of changing regulations**: AI laws are evolving quickly. Proactively monitor developments and adjust your AI strategies.

❖ **Compliance dictates AI strategy**: Embed transparency, fairness, and security into every AI project to avoid legal complications and build trust.

❖ **Expect future regulatory shifts**: Tightened privacy laws, bias audits, and AI-specific rules are on the horizon, impacting how you operate globally.

MINDSHIFT EXERCISE #12:
AI Compliance Checklist

Review your organization's compliance with AI-related regulations (e.g., GDPR, CCPA). Answer these questions:

- ❖ Is your AI use compliant with current data privacy laws?

- ❖ Do you have proper documentation for how AI decisions are made and stored?

- ❖ How are you staying up-to-date with evolving AI regulations?

Action Step: Create an "AI Compliance Checklist" to ensure your organization regularly reviews and updates its compliance with legal and ethical AI standards.

By embedding compliance into your AI strategy, you'll not only meet regulatory requirements but also enhance trust and accountability, positioning your organization for long-term success.

Now that you've mastered the compliance landscape, it's time to focus on innovation. In the next chapter, we'll dive into Boyd's Law and why the speed of iteration is more important than perfection when adopting AI in your business

Section Four:
Driving Innovation and Human-Centered AI for Lasting Success

The future of AI isn't just about who can implement it the fastest—it's about who can innovate while keeping the human experience front and center. In this final section, we're going to explore how to lead with AI in a way that pushes the boundaries of what's possible, while staying true to the core of what makes your business successful: the people, the creativity, and the connections. AI is the tool, but innovation and human connection are what will ensure your long-term success. This is about setting your business up for the future—on your terms.

Chapter 13:

Why Speed of Iteration Matters in AI

In the AI race, the tortoise beats the hare. Consistent iteration trumps sporadic perfection every time.

In business, the mantra is often to aim for perfection. But when it comes to AI adoption, perfection can be your biggest roadblock. Technology evolves faster than ever and waiting for the "perfect" AI system or solution is a recipe for falling behind. Enter Boyd's Law. John Boyd, a legendary military strategist, didn't just leave us with tactics for the battlefield—he left us with an invaluable principle for modern innovation: **speed of iteration** trumps perfection every time.

Think of AI like a race where the terrain changes constantly. The competitors who adapt the fastest, test, tweak, and improve in real time will always finish ahead of those waiting for the "perfect" moment. The secret isn't in flawless execution; it's in moving quickly, learning as you go, and evolving with every step.

The temptation to wait for the perfect solution is real. After all, AI feels like a major leap—it's new technology, new processes, and new risks. But here's the reality: waiting for perfection is a guaranteed way to lose ground.

Let's break down why:

1. **AI is evolving at lightning speed.** The tools you're using today might be outdated in six months, maybe even sooner. If you're constantly chasing perfection, you'll find yourself stuck in development while the world moves on without you. Adopting AI is about staying agile—embracing the

learning curve and moving quickly, even when things aren't perfect.

2. **Iteration fuels innovation.** Each version of your AI solution teaches you something new. Maybe the first version isn't perfect, but every tweak, every test, and every failure pushes you closer to a breakthrough. Innovation doesn't happen in a vacuum—it happens in the fast-paced rhythm of trial and error, learning, and refinement.

3. **Perfection is expensive—speed is resourceful.** Let's be honest: chasing perfection burns through time, money, and energy. Meanwhile, quick iterations and improvements allow you to make real progress without overextending resources. You don't need the perfect system on day one— what you need is something that works and can evolve with your needs.

The businesses that succeed with AI aren't the ones that waited for the perfect solution. They're the ones that embraced the messy, fast-paced process of iteration—adapting quickly, learning from each mistake, and staying agile enough to evolve as the technology advances.

Now, let's talk about how you can apply Boyd's Law to your AI strategy and ensure your business reaps the rewards of speed over perfection.

Start Small, Scale Fast

One of the most powerful strategies in AI adoption is starting small. You don't need to launch an organization-wide AI overhaul overnight. Instead, find manageable projects where AI can make an immediate impact—automating repetitive tasks, improving customer service interactions, or optimizing a specific workflow. Small wins are your fuel for scaling up.

These initial experiments give you the freedom to iterate quickly. By starting with smaller projects, you're not just limiting risk—you're setting the stage for learning, improvement, and scaling when the time is right. And once you've refined your approach with those small projects, you can extend those learnings across the rest of the organization.

Launch Fast, Learn Faster

This is the heart of Boyd's Law: stop waiting for perfection. Launch your AI tools or solutions before they're flawless. Watch them in action, gather feedback, and continuously improve. When you put AI to work, you gather real-world data and insights that help you iterate more effectively.

Think about launching an AI-powered chatbot. Your first version might handle only the most basic customer queries. But the real magic happens after launch—when you collect data on how people interact with it, refine the responses, and layer in more complex capabilities over time. The chatbot doesn't need to be perfect from day one, but every iteration brings you closer to that goal.

Embrace Failure as Forward Progress

Here's a secret: the faster you fail, the quicker you can pivot. Failure isn't something to be feared—it's a step in the journey to success.

When it comes to AI, failure isn't just inevitable—it's essential. It's how you learn what works, what doesn't, and how to refine your strategy.

The key is to fail quickly and learn even faster. If something doesn't work, you'll know sooner rather than later. And each failure offers insight into what needs improvement. It's not about getting everything right on the first try—it's about making progress with each attempt.

Keep an Eye on the Future, Pivot in the Present

Speed of iteration isn't just about focusing on short-term gains. You've got to stay adaptable and keep an eye on long-term trends. The world of AI is evolving fast, and being able to pivot when new advancements appear is crucial.

Your business needs to be ready to evolve along with AI, so always stay informed on emerging trends and technologies. The combination of quick iteration and long-term vision will keep your business on the cutting edge, ensuring that your AI tools remain not only effective but forward-thinking.

Creating a Culture of Experimentation

For Boyd's Law to truly take root in your organization, you need a culture that embraces experimentation. In traditional business settings, the fear of failure can paralyze innovation. But in the world of AI, experimentation—and, yes, even failure—is essential for success.

Empower your teams to experiment with AI. Whether it's marketing, HR, or sales, encourage every department to test new AI-driven strategies. The freedom to fail—and learn from those failures—will accelerate your AI adoption and fuel innovation across the board.

And remember to celebrate the small wins. When an AI project succeeds—whether it's automating a single task or improving a small process—make sure everyone knows about it. Recognizing these wins will build momentum and inspire other teams to dive into AI with the same enthusiasm.

Finally, shift the focus from perfection to learning. When people stop worrying about getting everything right on the first try, they're more willing to take risks and push the boundaries of what AI can do for your business. This mindset is the key to driving AI success and staying ahead in an ever-changing landscape.

Quick Win:
AI Speed Dating

Here's a fun way to foster creativity and rapid thinking. Set up an "AI Speed Dating" session with your team.

Pair up employees and give them five minutes to come up with a wild AI idea for your business. Then switch partners, reduce the time to three minutes, and refine the idea. Keep switching and shrinking the time until they're down to 30-second pitches.

You'll be amazed at the innovative ideas that surface when creativity meets urgency. Plus, it's way more fun than another traditional brainstorming session.

Key Takeaways

❖ **Speed of iteration beats perfection:** Launch fast, learn quickly, and keep improving based on real-world data.

❖ **Start small and scale:** Begin with small AI projects that deliver quick wins, then build on that success.

❖ **Embrace experimentation and failure:** Failure is part of the process. Learn from it and move forward.

❖ **Measure, refine, and adapt:** Use feedback loops to continuously improve your AI systems, ensuring they stay relevant and effective.

MINDSHIFT EXERCISE #13:
The Iteration Accelerator

Choose a recent AI project or initiative and reflect on its development process:

- ❖ How quickly were you able to implement and iterate on the project?

- ❖ What steps could you have taken to shorten the iteration cycle?

- ❖ What role did perfectionism play in the process, and how could you have embraced faster iterations?

Action Step: Set up an "Iteration Accelerator" by defining shorter cycles for AI project development. Focus on small, rapid improvements rather than waiting for perfection

Iteration and agility are key to AI success, but choosing the right tools is equally critical. In the next chapter, you'll learn how to select the best AI tools for your business and when to leverage general AI models versus industry-specific solutions.

Chapter 14:

Choosing the Right AI Tools for Your Business

The key to AI success isn't having the most advanced tools — it's having the right tools for your unique challenges.

Before we dive into the next level of AI insight, a quick reminder: with your purchase of *AI MINDSHIFT*, you get exclusive access to the Companion Guide, video tutorials, AI app reviews, and some powerful bonuses—all available instantly at: ProfitRichResults.com/aibooktoolkit. If you made it this far in the book, you've seen me share this a hundred times so go get them.

Popular Large Language Models (LLMs):

The secret to AI success isn't in chasing the latest, most advanced tools—it's about choosing the right tools that directly address your business challenges. This is where understanding the difference between Large Language Models (LLMs) and specialized AI apps becomes essential.

How LLMs Form the Foundation for Industry-Specific AI Tools

Large Language Models (LLMs) are versatile and powerful, serving as the backbone for many of the AI apps dominating the market today. These models—like ChatGPT, Claude, or other well-known platforms—have broad capabilities and are easily adaptable to various industries. Whether you're automating customer service, generating content, or analyzing data, LLMs can provide immense value across different business functions.

But LLMs are just the beginning. The real magic happens when these general-purpose models are fine-tuned for specific industries. Let's break it down:

1. **Broad Knowledge Base**: LLMs come packed with a wide-ranging understanding of human language and context. Straight out of the box, they can handle basic tasks like answering inquiries or generating generic content.

2. **Customizable for Specific Needs**: Many specialized AI apps take the core abilities of LLMs and refine them with industry-specific data. For example, a healthcare AI app might be trained on medical records, understanding complex medical terminology and patient interactions. Similarly, in finance, LLMs are adapted to handle fraud detection, market analysis, or risk management.

3. **Rapid Deployment**: One of the greatest benefits of LLMs is their flexibility and scalability. Businesses can quickly deploy specialized solutions built on these models, reducing

development time and allowing for faster innovation in competitive environments.

LLMs provide a strong foundation, but with thousands of apps being developed on top of these models, the challenge is figuring out which tool actually fits your business needs. Let's walk through when it makes sense to go with a general LLM and when it's smarter to opt for something more customized.

When to Use General AI Models vs Customized Solutions

So, how do you decide whether to stick with a general-purpose LLM or invest in a specialized AI solution for your business? Here's a quick guide to help you decide:

Use General AI Models (LLMs) When:

- **You Need Versatility**: LLMs can handle a variety of tasks, making them perfect for businesses that need flexibility. Whether you're automating customer service, generating reports, or creating marketing content, an LLM can juggle multiple roles with minimal setup.

- **You Want Speed**: If you're looking for quick deployment to test AI across different tasks, general-purpose LLMs are your go-to solution. They're ready out of the box, providing instant functionality while giving you room to fine-tune over time.

- **Your Use Case Is Broad**: For tasks that don't require specialized knowledge, an LLM can be more than enough.

Think simple automations, basic data analysis, or standard customer interactions that don't need deep industry expertise.

Use Specialized AI Solutions When:

1. **Your Industry Has Unique Needs**: If you're operating in highly regulated or complex industries—like healthcare, finance, or legal—you'll benefit from AI tools that are built specifically for your sector. These solutions come trained with the nuances, regulations, and standards that general LLMs might not understand.

2. **Precision Matters**: In sectors where accuracy is non-negotiable—think healthcare diagnosis or legal compliance—specialized AI apps will outperform general models. They're designed with industry-specific data to ensure better outcomes, tighter compliance, and reduced risk.

3. **You Need Deep Integration**: Often, specialized AI apps are built to work seamlessly with existing industry tools. For instance, a logistics company might need AI that can optimize shipping routes while integrating with their transportation management software. In these cases, general-purpose LLMs won't cut it.

With thousands of AI tools on the market and more launching every day, it's crucial to do your research. Test different tools, evaluate their strengths and weaknesses, and always make sure the app you choose solves a real problem for your business—whether that's improving workflows, enhancing decision-making, or boosting productivity. And as always, remember that AI should enhance, not replace, the human touch in your operations.

The Future of Multimodal AI Applications in Business

Now, let's take a look at the future. As AI technology evolves, we're seeing the rise of **multimodal AI applications**—AI systems that can process and interpret multiple types of data simultaneously. This is a game-changer. Imagine AI that can analyze text, images, audio, and video all in one seamless system.

Here's how multimodal AI is already reshaping business:

1. **Enhanced Customer Interactions**: With multimodal AI, businesses can create richer, more personalized customer experiences. Picture an AI that can interpret visual cues during a video call, understand tone in an audio interaction, and provide real-time text analysis—all at once. This will elevate customer service and allow for more dynamic, human-like interactions.

2. **Streamlined Operations**: In industries like logistics, healthcare, and manufacturing, multimodal AI can combine data from various sources—like sensor readings, video surveillance, and operational logs—to optimize decision-making. For example, a multimodal AI system in a factory could use video and sensor data to predict when machinery needs maintenance, avoiding costly breakdowns.

3. **Cross-Industry Flexibility:** The adaptability of multimodal AI means it can be used across different industries. A tool used for analyzing consumer behavior in retail might be just as useful for financial market analysis or media production. The future is all about flexibility and deeper insights.

The future of AI is all about integration. The ability to seamlessly combine data from different sources will lead to smarter decision-making, enhanced customer experiences, and more efficient business operations.

Quick Win:
AI Tool Showdown

Here's a practical exercise: set up an "AI Tool Showdown" for your team. Pick three different AI tools—a general-purpose LLM and two specialized AI apps.

Present your team with a real business challenge and give them 20 minutes to test each tool.

Afterward, have them rate the tools based on ease of use, quality of output, and potential business impact.

This hands-on experience will give you valuable insights into which tools are worth the hype and which ones actually deliver results. Plus, it's a great way to get your team excited about the potential of AI.

Key Takeaways:

❖ **LLMs form the foundation for many AI apps**: Large Language Models are flexible and powerful, making them a strong base for industry-specific AI tools that require fine-tuning. Start with ChatGPT, then expand to other AI Tools.

❖ **Do your research when selecting AI tools**: With thousands of AI apps out there, it's critical to test different solutions to find what truly meets your business needs.

❖ **Multimodal AI is the future**: The ability to integrate text, audio, video, and image analysis in one system will create smarter, more adaptive AI solutions for your business.

MINDSHIFT EXERCISE #14:
The AI Tool Selector

Evaluate the AI tools currently in use across your organization:

❖ Are you using general AI models when specialized tools would be more effective?

❖ Which AI tools are providing the most value, and which are underperforming?

❖ Are there emerging AI tools that could better meet your organization's needs?

Action Step: Create an "AI Tool Selector" matrix that helps you compare and evaluate new AI tools based on business fit, ease of integration, and potential ROI.

Now that you've learned how to select the best AI tools, how do you keep pushing forward with continuous innovation? In the next chapter, we'll explore how to build a culture of constant improvement and creativity with AI at its core.

Chapter 15:

Continuous Improvement and AI Innovation

With AI, complacency is the real competitor. Innovate continuously or watch your competitors do it for you.

Congratulations. Just a couple more chapters! You've already got a firm grasp on how AI can transform your business, but here's the next big truth: AI is not a "set it and forget it" tool. Just like a high-performing sports team, you don't stop practicing and refining just because you've got the best players. AI requires continuous adjustment, feedback, and an innovative mindset to keep it running at peak performance.

In this chapter, we'll explore how to build a continuous improvement culture into your AI strategy, from creating feedback loops that track success and highlight challenges, to fostering innovation through collaboration and internal competition. Plus, I'll remind you that if you want to take your AI game to the next level, I'm always a keynote or workshop away from helping your team elevate their AI strategy.

Creating Feedback Loops to Track AI Success and Challenges

You've heard it before—feedback is the lifeblood of improvement, whether it's from customers, employees, or systems. With AI, it's no different. You can't just launch AI into your business and assume it's working perfectly. To really leverage AI, you need constant data to evaluate its impact and course-correct where needed.

Here's how to build effective feedback loops into your AI strategy:

1. **Track Key Performance Metrics**: The first step is to establish performance indicators (KPIs) that align with your AI goals. Whether you're using AI for customer service, marketing, or supply chain management, you need to measure the right things. For customer service, track metrics like response times or customer satisfaction. For logistics, focus on inventory accuracy or process efficiency. If you're not measuring it, you can't improve it.

2. **Gather Employee and Customer Feedback**: While AI is often associated with automation, don't forget the human element. Your employees and customers are the ones interacting with these tools, so their feedback is critical. You might find that your AI chatbot is cutting wait times in half, but if customers find the responses robotic and unsatisfactory, that's a problem worth solving.

3. **Analyze and Iterate**: After gathering feedback, put it to use. AI systems aren't static—they need constant optimization. Maybe the AI is excelling in one area but struggling in another. This is where you pivot, tweak, or upgrade your models. Schedule regular check-ins—whether monthly or quarterly—where you evaluate the AI's performance and make adjustments as necessary.

Remember, continuous improvement isn't just a trendy buzzword. Without structured feedback loops, you're essentially flying blind, and in the fast-paced AI landscape, that's a disaster waiting to happen.

Promoting Innovation Through Internal AI Hackathons and Collaboration

One of the best ways to drive AI innovation within your business is by fostering a culture that promotes creativity and experimentation. A fantastic way to do that? **AI hackathons** and collaborative innovation sessions. And don't let the word "hackathon" scare you—this isn't just for the IT department.

Here's how to host an internal AI hackathon that can ignite new ideas across your organization:

- **Set Clear Objectives**: Whether it's improving customer experience, streamlining operations, or finding new ways to increase productivity, give your team a clear objective to solve with AI. Align the challenge with your business goals to keep it grounded in reality.

- **Cross-Functional Teams**: AI is most powerful when you bring people together from different departments—sales, marketing, operations, and HR all have valuable insights to offer. For example, your marketing team may have ideas on personalizing customer experiences with AI, while your HR team might find new ways to optimize recruitment. The best ideas often come from unexpected places.

- **Celebrate and Implement**: Don't let the excitement fizzle after the event. Celebrate the ideas your teams come up with and identify the ones with real business potential. Test them, iterate, and implement pilot programs to move forward. Whether it's a prototype or a full-fledged tool, what matters is that the momentum continues after the hackathon.

AI hackathons are a great way to drive innovation and reinforce a collaborative mindset around AI. Who knows, your next business breakthrough might come from that Friday afternoon AI hackathon you decided to host.

Sustaining a Culture of AI-Driven Improvement Over Time

Long-term success with AI isn't just about landing big wins—it's about building a culture that thrives on continuous improvement. AI is always evolving, and so should your approach to using it. Here's how to create a culture that's continuously pushing forward, even when the excitement of new technology wears off.

1. **Encourage Experimentation**: AI should never feel like a one-time project. Empower your teams to explore new ways AI can solve problems or improve processes. Let them experiment and, more importantly, let them fail. A culture that embraces risk-taking and learning from mistakes will naturally lead to more innovative solutions.

2. **Provide Ongoing Training**: AI technology changes fast, and if your team isn't keeping up, you'll quickly fall behind. Make ongoing AI training a priority. Whether it's through online courses, workshops, or bringing in an expert (like me!) for customized training sessions, keeping your team sharp is crucial for maintaining a competitive edge.

3. **Recognize and Reward Improvement**: Small wins matter. Recognize team members who have contributed to improving your AI tools or processes. Celebrate those innovations—whether it's more efficient workflows, improved customer interactions, or creative solutions that emerged from internal experimentation. When people feel their efforts are valued, they'll continue pushing for more.

Building a culture of continuous AI-driven improvement means staying adaptable, motivated, and always looking for the next creative way to use AI to drive growth. You want your team to see AI as a constantly evolving tool, not a static piece of technology.

Quick Win:
AI Improvement Jam

Here's a fun way to get your team involved in the process: Host an "AI Improvement Jam." Pick an AI tool or process your business is currently using, and give your team 30 minutes to brainstorm ways to improve it.

Encourage crazy ideas—sometimes the most out-of-the-box suggestions lead to breakthrough innovations. After the session, vote on the top ideas and put the best ones into action.

Not only will this generate improvements, but it will also reinforce the mindset that AI is a tool to be continuously refined.

Key Takeaways:

❖ **Feedback loops are essential:** Build systems to track AI performance, gather feedback, and make data-driven adjustments.

❖ **Foster innovation through collaboration:** Internal hackathons and cross-functional teamwork drive new AI solutions and keep your business on the cutting edge.

❖ **Sustain a culture of improvement:** Encourage experimentation, provide ongoing AI education, and celebrate incremental wins to keep innovation alive.

MINDSHIFT EXERCISE #15:
AI Innovation Hackathon

Organize an internal AI hackathon focused on innovation and continuous improvement:

- ❖ Identify a specific business problem or area for improvement.

- ❖ Split teams into groups and ask them to brainstorm innovative AI-driven solutions.

- ❖ Have each group present its solution, including steps for implementation and projected outcomes.

Action Step: Hold quarterly "AI Innovation Hackathons" to promote continuous improvement, collaboration, and experimentation across teams.

We've covered a lot of ground on how to keep your AI strategy sharp and adaptable. But here's a final reminder—AI is a tool, and it's how you use it that sets your business apart. If you want to take your AI strategy to the next level, I'm here to help, whether through a keynote, workshop, or more hands-on training for your team.

Visit ProfitRichResults.com to watch my speaker reel or learn more about fees and availability.

Innovation is important, but in a tech-driven world, maintaining the human touch is what will keep your business thriving. In the next chapter, we'll explore how to balance AI adoption with maintaining genuine human connections that build trust and loyalty.

Chapter 16:

Moving Forward– Maintaining the Human Touch in a Tech-Driven Future

The ultimate AI strategy isn't about replacing humans with machines; it's about creating a superhuman synergy between both.

As we near the end of this journey, we've explored how AI can transform your business—from streamlining operations to enhancing decision-making and managing crises. But as you move forward into an AI-driven future, one essential truth remains: no matter how advanced AI becomes, the human touch is irreplaceable. **To truly harness AI's power without losing what makes your brand and team special, you need to create a synergy between human connection and AI innovation.**

It's not about AI taking over—it's about AI amplifying what your team already does best. In this chapter, we'll dive into why the human touch will always be essential, how to embrace AI without losing authenticity, and how human-AI synergy is the future.

Why the Human Touch Will Remain Essential

AI is undoubtedly reshaping industries, communication, and decision-making processes. However, at the end of the day, people want to interact with other people. Customers crave authenticity, trust, and relationships, and your employees want to feel understood and valued beyond data metrics.

Here's why the human touch will always matter:

1. **Empathy Can't Be Automated**: AI can analyze sentiment, but it can't genuinely empathize. When a customer is

frustrated, or an employee is going through a rough patch, they want understanding, not just a solution. Empathy builds trust, and trust is the foundation of meaningful relationships, whether it's with customers, employees, or business partners.

2. **Creativity Thrives on Human Interaction**: While AI can generate ideas and even content, creativity is rooted in collaboration, emotional intelligence, and the exchange of diverse perspectives. The most innovative breakthroughs occur when humans bounce ideas off one another, challenge assumptions, and bring their unique insights to the table. AI can assist in the creative process, but it cannot replace the human spark that drives true innovation.

3. **Relationships Build Loyalty**: In a world of AI-driven automations, the businesses that stand out will be those that maintain personal, authentic relationships with their customers. AI can help streamline interactions, but it's the human touch that will make those interactions memorable and foster loyalty. People want to feel understood and valued—and that requires real human connection.

Steps for Leaders and Teams to Embrace AI Without Losing Authenticity

So, how can you harness AI's power without losing what makes your business human? The key is to use AI as a tool that enhances what you do, not a replacement for your unique strengths.

Here's how you can do it:

- **Integrate AI as an Assistant, Not a Replacement**: AI should automate routine tasks, provide data-driven insights, and help with decision-making—but the complex, emotional, and creative work should stay in human hands. Think of AI as a tool that helps your team be more efficient and impactful, not as something that replaces them.

- **Prioritize Customer and Employee Experience**: Before deploying any AI solution, always ask, *How does this improve the experience for my customers or employees?* If AI is being implemented but creating feelings of disconnection or frustration, it's counterproductive. Keep the human experience at the core of every AI initiative.

- **Foster Transparency and Communication**: Be transparent with your employees and customers about how you're using AI. People need to trust that AI is there to enhance their experience, not replace their value. Open, clear communication helps avoid fear of replacement and builds trust in the technology's role.

- **Train Your Team to Work with AI**: AI literacy is essential. Ensure your teams are trained not only to use AI but to understand that their judgment, creativity, and empathy are still critical. AI can process data, but it's humans who make decisions based on context, insight, and intuition. Regular training keeps everyone up to speed on how to leverage AI effectively while maintaining their role as the heart of your business.

By following these steps, you can fully leverage AI's potential without losing the core human elements that make your company, brand, and team exceptional. It's about leading with humanity and using AI to amplify your strengths rather than overshadow them.

Final Thoughts on Future AI Trends and Human-AI Synergy

As we look toward the future, AI will continue to evolve rapidly. We'll see advancements in **multimodal AI**, where systems can seamlessly integrate text, video, audio, and images into their decision-making processes. AI will become more deeply embedded in business operations, from real-time customer service support to predictive analytics that help drive strategic decisions.

But one thing will remain constant: the need for **human-AI synergy**. The businesses that succeed in the AI-driven future will be those that know how to combine AI's efficiency and scalability with human creativity, empathy, and leadership.

The future won't be about choosing between AI or humans—it's about combining the best of both. AI will help process information, scale operations, and handle repetitive tasks, while human intelligence will bring the creativity, emotional intelligence, and decision-making that will drive your business forward.

We'll also see more AI tools designed to promote **collaboration between humans and machines**, making it easier for leaders to make informed decisions and innovate in ways we haven't yet imagined. The future is about creating a powerful blend where both AI and human strengths are amplified.

Quick Win:
Human Touch Checkpoint

Create a "Human Touch Checkpoint" for your AI-driven processes. Choose one AI-powered customer interaction (such as a chatbot conversation or automated email) and one internal AI tool your team uses regularly.

Identify three key points where adding a human element could enhance the experience. Maybe it's a personalized follow-up message after a chatbot interaction or a brainstorming session to complement an AI-generated report.

Test these human touchpoints for a week and gather feedback. You'll likely find that even small human interventions can make a significant difference.

Key Takeaways:

❖ **The human touch remains irreplaceable**: AI can automate tasks and enhance operations, but it can't replicate empathy, creativity, or meaningful relationships.

❖ **Use AI to enhance, not replace**: Leaders should leverage AI to automate routine tasks and provide decision-making insights, while maintaining the personal, authentic aspects of their brand and team culture.

❖ **Transparency builds trust**: Be open with your employees and customers about how and why you're using AI. Transparency keeps the human element at the forefront and ensures AI is seen as a supportive tool, not a replacement.

❖ **The future lies in human-AI synergy**: The businesses that will lead the next wave of innovation are those that effectively blend AI's strengths with human insight, creativity, and leadership.

MINDSHIFT EXERCISE #16:
Human-AI Synergy Mapping

Map out how AI and human interactions currently coexist within your organization:

❖ Where is AI driving efficiency, and where are humans providing the emotional intelligence needed for personal connection?

❖ Are there opportunities to better integrate AI while preserving human touchpoints?

❖ How can you ensure that AI enhances, rather than diminishes, authenticity and trust in customer and employee relationships?

Action Step: Create a "Human-AI Synergy Map" that outlines how AI and human interactions are balanced. Use this map to guide decisions on future AI implementations.

As we wrap up this chapter, remember that AI's true potential is realized when it works in harmony with human creativity, empathy, and leadership. The businesses that will succeed in this new landscape aren't the ones with the most sophisticated AI, but those that combine AI's capabilities with the irreplaceable human touch.

But this is just the beginning. In the next and final chapter, *AI Mindshift Action Plan – Your Next Steps*, we'll lay out a roadmap to turn all these insights into concrete actions.

It's time to build on everything you've learned and take decisive steps toward integrating AI into your business in a way that drives growth and keeps your human core intact.

Chapter 17:

The AI Horizon: Emerging Trends and Their Business Impact

The future of AI isn't just coming—it's already here, evolving at breakneck speed. The businesses that anticipate and adapt will be the ones shaping tomorrow's market.

This chapter isn't about predicting the future—it's about preparing for it. The trends we'll explore will show you where I think AI is headed, but more importantly, how you can leverage these advancements to future-proof your business. And yes, I do have a crystal ball on my desk, but the real key here is strategic foresight.

Multimodal AI: Talk, Text, Images, Video

Large Language Models (LLMs) are a hot topic in AI, but the future of AI is about much more than generating text. The next big step is multimodal AI—technology that combines text, voice, images, and even touch. This allows businesses to gather and respond to data from many sources at once, opening up new possibilities for smarter, more interactive applications.

Trend 1. Multimodal AI: The New Business Sixth Sense

Multimodal AI will revolutionize how businesses understand and interact with their environments by integrating different data sources:

- **Customer Insights**: Analyze facial expressions during video calls while processing tone of voice and conversation content to gain deeper insights.

- **Quality Control**: In manufacturing, AI will combine visual and acoustic analysis to detect defects invisible to the human eye.

- **Immersive Marketing**: Create personalized, interactive ad experiences that adapt in real time based on viewer engagement.

Trend 2. Edge AI: Bringing Intelligence to the Front Lines

AI is also moving closer to the action with **Edge AI**, which processes data on local devices rather than in the cloud. This shift allows businesses to make real-time decisions with greater security and personalization.

- **Real-time Decision Making**: AI-powered devices will make split-second decisions on-site without relying on cloud connectivity.

- **Enhanced Privacy**: Local data processing keeps sensitive information closer to home, addressing growing privacy concerns.

- **Personalized Experiences**: Deliver hyper-personalized services based on real-time, local data, improving both customer satisfaction and operational efficiency.

Trend 3. AI and Quantum Computing: A Game-Changing Synergy

The combination of **AI and quantum computing** promises to

tackle complex challenges at incredible speeds, unlocking possibilities we once thought were science fiction.

- **Supply Chain Optimization**: AI-quantum systems will solve logistical challenges that currently take hours in a matter of seconds, saving millions in operational costs.

- **Drug Discovery**: These systems will speed up drug development, revolutionizing healthcare.

- **Financial Modeling**: Hyper-accurate predictive models will transform industries by refining risk assessment and investment strategies.

Trend 4. AI-Human Collaboration 2.0: The Rise of AI Assistants

The next generation of AI assistants will go beyond basic chatbots. These systems will work seamlessly with humans to tackle complex problems, evolving into indispensable business tools.

- **AI Middle Managers**: AI systems will break down business goals into actionable tasks and manage both human and AI teams to accomplish them.

- **Creative Collaboration**: AI will work alongside humans to enhance creativity, from generating ideas to helping execute them in real time.

- **Continuous Learning**: AI assistants will learn from every interaction, becoming smarter and more adaptable with each use.

Trend 5. Ethical AI: The Non-Negotiable Future

As AI becomes more powerful, **ethical considerations** will move from the periphery to the center of AI strategy. Businesses that focus on ethical AI now will set themselves up for long-term success.

- **Regulatory Compliance**: AI-specific regulations are coming fast. Businesses that prioritize ethical AI will stay ahead of the curve, avoiding fines and legal trouble.

- **Transparency and Explainability**: The "black box" nature of AI will become increasingly unacceptable. AI systems must be able to explain their decisions clearly.

- **Bias Detection and Mitigation**: AI bias detection will become a standard practice as businesses strive to create fair, unbiased systems.

Trend 6. AI-Generated Content and Deep Fakes: Blurring the Lines of Reality

AI tools for generating text, images, audio, and video are evolving rapidly, making it nearly impossible to distinguish between human-created and AI-generated content. This trend brings both opportunities and challenges.

- **Hyper-Realistic Content Creation**: AI will allow businesses to create personalized content at an unprecedented speed and scale, revolutionizing marketing campaigns.

- **Deep Fakes and Synthetic Media**: These technologies will extend beyond entertainment, finding uses in training, customer service, and personalized marketing.

- **Authentication Challenges**: As AI-generated content becomes more prevalent, verifying authenticity will be crucial. Expect AI-powered authentication tools and blockchain-based verification systems to rise in importance.

Innovate or Fall Behind: The Stakes Are High

The businesses that thrive in the coming decade won't just use AI—they'll pioneer new applications, set ethical standards, and lead entire industries into the future. Standing still isn't an option. The question is: Will you lead the charge or be left playing catch-up?

The pace of AI development isn't slowing down, and the companies that fail to adapt risk stagnation or, worse, becoming obsolete. Those that embrace these emerging trends, however, will find themselves in a position to not only survive but thrive in a rapidly changing marketplace.

Quick Win:
Future-Proofing Brainstorm

Bring your team together for a 30-minute brainstorming session.

Pick one of the emerging AI trends from this chapter and discuss how it could impact your business.

Come up with at least three potential applications and one challenge you might face.

This exercise will help you kick-start your strategic planning and prepare for the future of AI.

Key Takeaways

❖ **Multimodal AI** will unlock unprecedented insights by combining multiple data sources.

❖ **Edge AI** will bring real-time intelligence to the forefront of operations, improving privacy and personalization.

❖ **Quantum-AI integration** will solve complex problems at incredible speeds, transforming industries from logistics to healthcare.

❖ **Next-gen AI assistants** will take collaboration to new heights, working seamlessly with human teams.

❖ **Ethical AI** will become a business imperative, with transparency and fairness at the forefront.

MINDSHIFT EXERCISE #17:
Trendspotting with AI

❖ Look at your industry and consider emerging AI trends that could impact your business:

❖ Which trends are most relevant to your organization's future success?

❖ How can you position your business to take advantage of these trends early?

❖ What internal adjustments will be needed to capitalize on these opportunities?

Action Step: Create a "Trendspotting AI Watchlist" to monitor developments in AI technology and assess how they might impact your business over the next 12 to 18 months.

As we wrap up this glimpse into the future, remember: the most successful businesses won't just adapt to these trends—they'll shape them.

In our final chapter, we'll translate all we've learned into an actionable plan to position your business at the forefront of the AI revolution.

Chapter 18:

AI Mindshift Action Plan – Your Next Steps

An AI Mindshift isn't just a change in thinking — it's a call to action. Your future success is built on the steps you take today.

You've made it through 17 chapters, diving deep into the transformative power of AI. Now, it's time to turn all that insight into action. This final chapter isn't just a recap; it's a roadmap—a clear, concise action plan to help you think about, implement, adapt, and continuously improve AI in your business.

Step 1: Think About AI Strategically

AI success isn't about chasing the latest tools—it's about aligning AI with your business goals to create real value.

1. **Adopt an AI-Critical Thinking Mindset**: Always ask yourself how AI can directly solve business challenges. Don't just implement AI for the sake of it—make sure it addresses a specific pain point, whether that's automating repetitive tasks, improving decision-making, or enhancing customer experiences.

2. **Maintain the Human Touch:**: As you've learned, AI is powerful, but it should enhance, not replace, the personal and human elements of your business. Continuously ask, *How does this improve the customer or employee experience?* Never let the technology overshadow the authenticity that builds trust and loyalty.

Quick Win: Host an **"AI Opportunity Brainstorm"** with your team. Identify your top three challenges and explore how AI might address each one, always keeping the human element in mind.

Step 2: Implement AI with Purpose

Implementing AI requires thoughtful integration into your operations—it's not just about downloading an app. Here's how to implement AI effectively:

1. **Start Small and Scale**: Don't overwhelm your organization by trying to roll out AI across every department all at once. Start with small, focused projects that deliver immediate impact, such as automating customer support or improving lead scoring. Once you see success, scale those projects to other areas of your business.

2. **Choose the Right Tools**: There are thousands of AI apps out there, but don't get distracted by the latest shiny tech. Do your due diligence and choose tools that solve specific problems, improve workflows, and boost productivity—while preserving the human connection that defines your brand.

3. **Empower Your Team**: AI is a tool, but your people are still your greatest asset. Invest in AI training to ensure your teams know how to use these tools effectively, always keeping customer experience and business goals at the forefront.

Quick Win: Identify one repetitive task that consumes time. Implement an AI solution to automate it within the next 30 days, and track time saved as a case study for future AI adoption.

Step 3: Adapt, Iterate and Optimize

AI is always evolving, and your approach should be too.

1. **Establish Feedback Loops::** Regularly track AI performance by gathering feedback from customers and employees. This will help you fine-tune your AI systems to ensure they align with your business objectives..

2. **Embrace Experimentation**: Foster a culture where trying new AI-driven solutions and learning from fast failures is encouraged. Hold internal AI hackathons, promote cross-functional collaboration, and stay open to innovation. This culture is key to long-term success.

Quick Win: Institute a monthly "**AI Learning Hour**" where team members share new AI tools they've discovered, lessons learned from implementations, or insights from AI industry news.

Step 4: Commit to Continuous Improvement

Staying ahead in AI means more than keeping up—it means consistently improving and pushing boundaries.

1. **Promote a Culture of Innovation**: Encourage your teams to always look for new AI solutions that can enhance workflows, decision-making, or customer experiences. Empower everyone to contribute ideas and reward experimentation.

2. **Prioritize Ethical AI**: Make sure your AI initiatives align with your values and meet all regulatory standards. Develop an "AI Ethics Checklist" to review each project before launch

3. **Stay Connected and Informed**: AI is constantly evolving, and so should your knowledge. Connect with me, Ford Saeks, on LinkedIn at linkedin.com/in/fordsaeks for regular updates. Watch my latest AI Video Training at ProfitRichResults.com/ai-training

Quick Win: Schedule a quarterly "**AI Performance Review**" to assess what's working, what needs tweaking, and where to go next.

Step 5: Engage and Share Your Experience

1. **Leave an Amazon Review**: If this book has provided value, please take a moment to share your thoughts by leaving an Amazon review. Not only does your feedback help other readers, but it also supports the ongoing conversation around AI and business growth.

2. **Check Out the Resources and Connect**: I've included a list of recommended resources in the back of the book. Dive into those to continue expanding your AI knowledge. Let's keep the conversation going—I'd love to hear your thoughts and continue sharing the latest in AI.

Quick Win: Take a moment to leave an Amazon review, then dive into the resources section to continue your AI journey. And seriously, connect with me on LinkedIn.

Step 6: Measure Success and Celebrate Wins

Tracking progress and recognizing achievements will keep momentum high and ensure your AI initiatives are driving real value.

1. **Define Key Metrics:** Identify clear metrics to evaluate the impact of AI on your business. These could include time savings, cost reductions, customer satisfaction improvements, or increased productivity. Regularly assess these metrics to measure success and make informed adjustments

2. **Celebrate Small Wins:** Recognize and reward your team's AI milestones, no matter how small. Celebrating achievements builds morale and reinforces the positive impact AI is having on your business.

3. **Share Results Across Teams:** Transparency fuels progress. Share the results and lessons learned from AI projects across departments to foster collaboration, inspire new ideas, and maintain alignment.

Quick Win: Host a quarterly "**AI Impact Review**" where you analyze key metrics, celebrate successes, and identify areas for improvement. Recognize team contributions and share progress to keep everyone invested in your AI journey.

Last Easer Egg... if you want to jump on a quick zoom call with me visit www.15withford.com

MINDSHIFT EXERCISE #18:
The AI Implementation Roadmap

Develop a personal action plan based on the key lessons in *AI MINDSHIFT*:

- ❖ What are the top three AI initiatives you will implement within the next 90 days?

- ❖ Who on your team will help drive these initiatives forward?

- ❖ What metrics will you use to measure success?

Action Step: Build an "AI Implementation Roadmap" that outlines your next steps for adopting and optimizing AI across your organization.

Revisit this roadmap every quarter to track progress and adapt as necessary.

Final Thought: Prioritize Action Over Perfection

The most important lesson to take away is this: **action beats perfection every time**. AI is evolving too quickly to wait for the perfect plan or the ideal tool. Companies that succeed aren't the ones holding out for perfection—they're the ones taking decisive steps forward, learning, and iterating along the way.

By following the action plan laid out in this book, you'll unlock AI's potential while preserving the human touch and authenticity that set your company apart. Whether you're just beginning your AI journey or scaling up, the key is to remain agile, innovative, and focused on delivering value to your customers.

Now is the time to act. The future of AI is here, and it's those who move first and adapt that will lead the way. Keep pushing, keep learning, and keep refining—because there's no better moment to elevate your AI strategy than right now. Your AI-powered future starts today.

Next Steps: Let's Work Together

If you're ready to put these ideas into action, let's make it happen together. Whether you need a keynote speaker to motivate your team, an interactive workshop to sharpen AI skills, or consulting tailored to your business challenges, I'm here to help. For the almost the time in this book visit ProfitRichResults.com to explore how we can take your business to the next level.

The future of AI is calling, and you're in the perfect position to lead the charge.

Recommended
Resources

Keynote Presentations and Training Solutions.

BRIDGE THE GAP, GROW YOUR BUSINESS.

Ford Saeks, a Hall of Fame Keynote Speaker and Business Growth Accelerator, delivers dynamic presentations to help organizations elevate performance, accelerate growth, and generate profits. He has partnered with corporations and associations worldwide, from start-ups to billion-dollar brands, providing actionable insights for bridging gaps in mindset, strategy, and tactics.

Unleashing AI for Every Role

In today's rapidly evolving business landscape, AI is more than just a technological advancement—it's a game-changing tool that can revolutionize how organizations operate. Ford Saeks is at the forefront of helping businesses harness AI to unlock

185

unprecedented efficiency and productivity across all roles—from HR and Operations to Marketing and Sales. He offers tailored AI training programs that provide immediate, practical results, helping teams streamline tasks, craft effective AI prompts, and integrate AI seamlessly into daily operations.

Ford's in-demand keynote presentations and consulting services focus on demystifying AI, ensuring that companies not only adapt to change but thrive in an AI-driven world. His hands-on, interactive sessions leave participants with actionable insights to immediately implement AI and gain a competitive edge in their respective industries.

On-Stage, On-Screen, On-Demand – Customized Keynotes and Workshops

Ford's Most Requested Presentations:

- **Business Growth Acceleration:** Maximize Your Findability, Accountability, and Profitability
- **Superpower Success:** Unleash Your Inner Superhero to Bust Through Barriers, Ignite High Performance, and Boost Results.
- **Winning Workplace Strategies:** How to Find, Engage, and Retain Top Talent
- **Unleashing the Power of AI for Any Role in Your Organization:** What You Need to Know Now to Rule Your World
- **Innovative Marketing & Sales Mastery:** Leveraging Your Brand to Build Trusted Relationships & Skyrocket Your Sales

- **Remarkable Customer Experiences:** Creating a Customer-Centric Culture That Drives Repeat & Referral Sales

Discover how Ford Saeks can help your organization master AI adoption and gain long-term success by watching his AI topic teaser video or speaker demo at www.ProfitRichResults.com

Give us a call at 1-800-946-7804 or 316-844-0245 or online at www.ProfitRichResults.com

Watch Ford's Keynote Speaker Trailer

On-demand AI Video Training

Take control of your business's future with Ford Saeks' comprehensive, on-demand AI video training. Designed for business leaders, marketers, and professionals across all industries, this training equips you with the latest AI tools and strategies. With frequently updated content and regular Zoom webinars, you'll always have access to cutting-edge insights and live interaction with Ford. Whether you're a beginner or advanced user, these resources are tailored to help you achieve immediate results.

What You'll Gain:

- **Up-to-date AI Strategies**: Stay on the leading edge of AI advancements with regularly refreshed content.

- **Practical Tools & Techniques**: Learn step-by-step methods for integrating AI into your daily operations, marketing, and sales processes.

- **Immediate ROI**: Implement actionable insights as soon as you finish the training to boost productivity and streamline workflows.

- **Tailored for All Roles**: Whether you're in HR, sales, or management, discover AI applications specific to your department.

- **Real-Life Case Studies**: Learn from successful AI implementations to see how other businesses are leveraging this technology for growth.

- **Expert Guidance**: Ford Saeks brings decades of business and AI experience to deliver training that is relevant, engaging, and results-driven.

- **Live Webinars**: Gain access to Ford's regular Zoom webinars for live, interactive training sessions. Catch the latest one at ProfitRichResults.com/ai-training.

Stay ahead in this fast-moving, AI-driven world with on-demand training and live webinars that deliver immediate value to your business. Visit the link to get started today!

Stay Ahead with the Latest AI Tools

In the fast-paced world of AI, staying up to date is crucial for success. For the latest and most effective AI tools that can revolutionize your business, visit ProfitRichResults.com/ai-tools. This page is constantly updated with cutting-edge tools and resources to help you leverage AI for marketing, sales, and productivity as the technology rapidly evolves.

Superpower: A Superhero's Guide to Leadership, Business and Life - Book

Superpower! is your guide to taking personal accountability for your leadership, business, and life. Through practical strategies, you'll learn how to rethink your mindset, reframe your goals, and refocus on the instincts that drive success. This handbook empowers you to:

- Develop a clear vision and action plan.

- Maximize your strengths and simplify your daily tasks.

- Use time-saving tools to boost productivity.

- Cultivate gratitude and learn from past experiences.

Packed with actionable steps and inspiring stories, *Superpower!* provides the tools to bridge the gap between where you are now and where you want to be. Unlock your potential and master your destiny—grab your copy today on Amazon or your favorite book reseller!

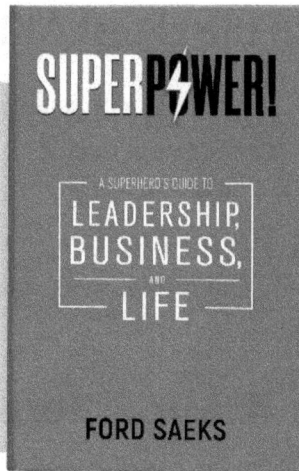

ACCELERATE: The Ultimate Franchisees Guide to Maximize Local Marketing and Boost Sales - Book

Are you a local franchise owner struggling with marketing, sales, and customer retention? *ACCELERATE* is your go-to guide for reigniting your local area marketing and sales efforts while staying aligned with your brand's standards. Whether you're an owner-operator juggling multiple roles, a franchise executive supporting your network, or a top producer looking to stay competitive, this book delivers practical insights to help you:

- Boost visibility, engagement, and conversions.

- Build a strong team and foster customer loyalty.

- Comply with brand guidelines while thriving locally.

Packed with actionable strategies, *ACCELERATE* equips you with the tools to dominate your market, improve customer experiences, and drive sustainable growth. Stop struggling and start seeing results— grab your copy today on Amazon or your favorite book reseller

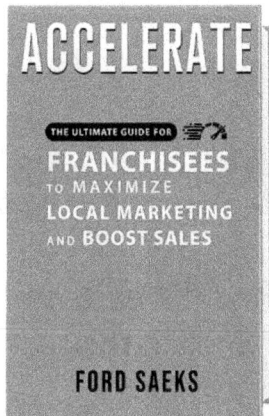

Connect with Ford Saeks Online

LISTEN: Listen to *The Business Growth Show* with Ford Saeks on your favorite streaming platform where we explore the latest strategies and tactics for growing your business and achieving success. Tune in each Wednesday at 11AM CST (or listen anytime) for valuable takeaways to help you accelerate your business's growth. Let's get growing!

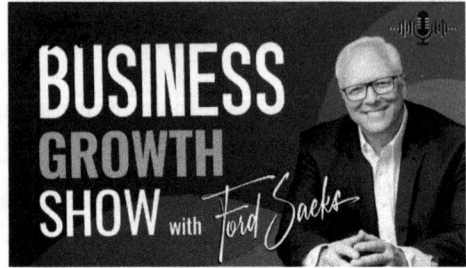

WATCH: *FORDIFY LIVE!* Weekly Episodes and more on YouTube, subscribe at www.Fordify.TV

DISCOVER:
www.ProfitRichResults.com

CREATIVE AGENCY:
www.PrimeConcepts.com

LINKEDIN:
www.LinkedIn.com/in/fordsaeks

TWITTER:
www.Twitter.com/prime_concepts

FACEBOOK:
www.Facebook.com/profitrichresults

INSTAGRAM:
www.Instagram.com/fordsaeks

TIKTOK:
www.tiktok.com/@fordsaeks

60 Ways to Put AI to Work in Your Business

Operations & Workflow

1. Automated email inbox management and smart response drafting

2. Converting existing processes into clear Standard Operating Procedures

3. Intelligent meeting scheduling and follow-up automation

4. Meeting transcription and key point summarization

5. Dynamic project timeline creation and adjustment

6. Automated expense report processing and categorization

7. Business report generation from raw data

8. Employee training material creation and updates

9. Smart document review and contract analysis

10. Cross-platform data entry automation

Marketing Impact

1. AI-powered social media content calendar creation

2. Multi-variant marketing copy generation for different audiences

3. Competitor social media strategy analysis

4. Blog post ideation and outline generation

5. Personalized email marketing campaign design

6. Website content SEO optimization

7. Social media image concept generation

8. Customer feedback and review analysis

9. Platform-specific ad copy creation

10. Industry trend identification and content planning

Sales Growth

1. Personalized sales outreach email scaling

2. Custom sales proposal generation

3. Sales call analysis and improvement insights

4. Lead scoring and prioritization

5. Converting follow-up email sequence creation

6. Quick sales presentation development

7. Sales conversation summarization

8. Customer purchase prediction modeling

9. Customized objection response creation

10. Automated CRM updates from conversations

Customer Experience

1. Basic chatbot implementation for common questions

2. Smart customer inquiry routing

3. Personalized customer appreciation messages

4. Quick customer email response generation

5. Customer feedback analysis and insight extraction

6. Effective survey design and analysis

7. Knowledge base development from customer interactions

8. Proactive customer issue prevention

9. Smart product recommendation system

10. Automated customer re-engagement campaigns

Data & Technical Solutions

1. Spreadsheet data cleaning and organization

2. No-code data visualization creation

3. Key information extraction from documents

4. Raw data formatting and structuring

5. Automated business report generation

6. Simple database creation without coding

7. Website update automation

8. Common task code snippet generation

9. No-code automation workflow building

10. Business metric trend analysis

Quick Start Tips:

- Begin with one high-impact task from each category

- Start with non-critical processes

- Always review AI outputs before use

- Keep up with new AI tool releases

- Share successful implementations

Your Quick Reference Guide for AI Prompting Success

Quick Tips Before You Start

- **Have a clear goal in mind:** Define what you want the AI to accomplish.

- **Gather necessary information:** Ensure you have relevant data or context to guide the AI.

- **Allow time for iterations:** Expect multiple rounds of refinement to improve results.

- **Keep your first attempts simple:** Begin with basic instructions before layering complexity.

- **Have examples ready if needed:** Reference prior work or templates to guide the AI.

- **Use specific, actionable language:** Clear and concise prompts yield better responses.

- **Ask for clarification:** If the AI output isn't what you expected, prompt it with "Do you understand? If not, ask me clarifying questions until you do."

Pro Tips for Better Prompting Results

- Start with "I need help with..." or "My goal is..."

- Use bullet points for clarity

- Number steps or requirements

- Include examples where possible

- Keep sentences short and clear

Common Mistakes to Avoid

- Being too vague

- Providing too little context

- Skipping the verification step

- Making unstated assumptions

- Requesting too much at once

Power Phrases to Use

- "Please confirm you understand…"

- "What questions do you have…"

- "Key requirements include…"

- "Success looks like…"

- "Let's approach this step by step…"

Red Flags to Watch For

- Vague or generic responses

- Missing key requirements

- Misaligned tone or style

- Incorrect assumptions

- Incomplete outputs

Quick Fix Solutions

If Results Aren't Right:

1. Add more context

2. Break into smaller tasks

3. Clarify requirements

4. Provide examples

5. Ask for explanations

Before Submitting, Check:

1. Clear goal stated?

2. Context provided?

3. Requirements listed?

4. Format specified?

5. Understanding verified?

Remember:

- Quality In = Quality Out

- Verify Before Proceeding

- Start Simple, Then Expand

- Review and Refine

- Save Successful Prompts

Effective AI Prompt Examples for Business Tasks

Using these examples as templates for similar tasks

Note: When using these templates, replace bracketed text [like this] with your specific information.

Best Practices Demonstrated in These Prompts:

1. Clear context and background information

2. Specific deliverable requirements

3. Format and style preferences

4. Constraints and limitations

5. Verification of understanding

6. Audience consideration

7. Structured output requests

8. Examples or templates where needed

9. Success criteria

10. Permission to ask clarifying questions

Operations & Workflow Example

Task: Converting existing processes into Standard Operating Procedures

PROMPT: I need to create a detailed SOP for our customer return process. Here's our current process:

[Paste your current process here]

Please help me transform this into a clear SOP with the following:

- A clear title and purpose statement

- Step-by-step numbered instructions

- Required materials/systems needed

- Key definitions

- Common troubleshooting scenarios

- Related procedures/references

Format it using clear headers and bullet points for easy reading. Use simple, direct language suitable for new employees.

Do you understand what I'm asking for? If not, please ask questions until you do. Once you confirm understanding, please proceed with creating the SOP.

Marketing Impact Example

Task: Multi-variant marketing copy

PROMPT: I'm creating marketing copy for our new [product/service name], a [brief description].

Target audiences:

1. [Describe first audience segment]

2. [Describe second audience segment]

3. [Describe third audience segment]

For each audience, please create:

- A headline (max 10 words)

- Three bullet points highlighting key benefits

- A call to action (max 8 words)

Brand voice: [Describe your brand voice]

Must include: [Any required terms/messaging]

Must avoid: [Any terms/approaches to avoid]

Please vary the emotional appeals and value propositions for each audience segment while maintaining consistent brand messaging.

Do you understand what I'm asking for? If not, please ask questions until you do. Once you confirm understanding, please proceed with the copy variants.

Sales Growth Example

Task: Customized objection response

PROMPT: Help me create effective responses to our top sales objections for [product/service name].

Price point: [specify]

Target customer: [describe ideal customer]

Please provide response scripts for these common objections:

1. "It's too expensive"

2. "We're happy with our current solution"

3. [Add your specific objection]

For each response, please:

- Acknowledge the concern

- Provide value-based counterpoints

- Include relevant social proof

- Add a question to move the conversation forward

- Keep responses under 30 seconds when spoken

- Include both formal and casual versions

Do you understand what I'm asking for? If not, please ask questions until you do. Once you confirm understanding, please proceed with the response scripts.

Customer Experience Example

Task: Knowledge base development from customer interactions

PROMPT: I need to create knowledge base articles based on our customer support conversations. Here's a batch of recent customer inquiries:

[Paste 5-10 recent customer questions/issues]

For each common issue, please create:

- A clear, searchable title

- A brief overview (2-3 sentences)

- Step-by-step resolution steps

- Relevant screenshots placeholders

- Related articles suggestions

- Troubleshooting tips

- Tags for categorization

Use plain language and avoid technical jargon unless necessary. Format for both web and mobile reading.

Do you understand what I'm asking for? If not, please ask questions until you do. Once you confirm understanding, please proceed with the knowledge base articles.

Data & Technical Example

Task: Business metric trend analysis

PROMPT: I need to analyze our key business metrics for the past [timeframe]. Here's our data:

[Paste your data or describe its format]

Please analyze for:

1. Significant trends and patterns

2. Seasonal variations

3. Anomalies or outliers

4. Correlations between metrics

5. Growth rates and changes

Provide a summary of key findings (non-technical language)

- Specific actionable insights and Recommendations for improvement

- Suggested metrics to track in the future and Potential areas of concern

Format the analysis for presentation to [specify audience: executives, team members, etc.]

Do you understand what I'm asking for? If not, please ask questions until you do. Once you confirm understanding, please proceed with the analysis.

Action Items Checklist

1. **Identify AI Opportunities:** Review key business processes for AI integration opportunities, from customer service to data analysis.

2. **Test AI in a Key Area:** Implement AI in non-critical tasks to evaluate its impact and scalability.

3. **Leverage AI for Data-Driven Decisions:** Use AI to inform strategic decisions like market trends, pricing, and customer insights.

4. **Explore Multi-Platform Integration:** Use various AI tools for holistic solutions across marketing, sales, and operations.

5. **Implement Continuous Adaptation:** Stay updated on AI advancements and integrate emerging tools to maintain agility.

6. **Foster AI Literacy in Your Team:** Provide AI training for your staff to ensure successful adoption and use.

7. **Create AI Governance Policies:** Establish clear ethical guidelines to ensure responsible AI usage and data protection.

8. **Develop Custom GPT Models:** Build customized AI models for specific business needs by uploading knowledge sources.

9. **Conduct Regular AI Performance Reviews:** Continuously monitor AI performance and refine tools based on results and feedback.

10. **Book a Customized AI Presentation:** Contact me for a value-driven keynote tailored to your organization, filled with actionable AI strategies for growth and success.

AI Glossary

Artificial Intelligence (AI):
Technology that mimics human intelligence, enabling machines to perform tasks like decision-making and problem-solving.

Large Language Model (LLM):
An AI model trained on vast amounts of text data to generate human-like text.

Automation:
Using AI to perform repetitive tasks without human intervention.

Predictive Analytics:
AI-based analysis to predict future outcomes based on historical data.

Natural Language Processing (NLP):
AI's ability to understand, interpret, and respond to human language.

Machine Learning (ML):
A subset of AI where models learn from data to improve over time.

Chatbot:
An AI-driven tool that simulates human conversation, often used in customer service to handle inquiries and provide support.

Robotic Process Automation (RPA):
Technology that uses software robots to automate highly repetitive and routine tasks traditionally performed by humans.

Deep Learning:
A type of machine learning that uses neural networks with many layers (deep neural networks) to analyze various levels of data features.

Computer Vision:
AI technology that enables machines to interpret and make decisions based on visual inputs like images and videos.

Sentiment Analysis:
The use of NLP to determine the emotional tone behind a series of words, helping businesses understand customer opinions and feedback.

Business Intelligence (BI):
Technologies and strategies used to analyze business information, often enhanced by AI to provide deeper insights and predictive capabilities.

Cognitive Computing:
AI systems that simulate human thought processes to solve complex problems and enhance decision-making.

Recommendation Engine:
AI algorithms that analyze user behavior and preferences to suggest products, services, or content tailored to individual users.

Generative AI:
AI systems capable of creating new content, such as text, images, or music, based on the data they have been trained on.

Data Privacy:
The practice of protecting personal and sensitive information collected by businesses, ensuring compliance with regulations and maintaining customer trust.

Ethics in AI:
The study and implementation of moral guidelines to ensure AI technologies are used responsibly and do not cause harm.

Explainable AI (XAI):
AI systems designed to provide understandable explanations for their decisions and actions, enhancing transparency and trust.

Reinforcement Learning:
A type of machine learning where an AI agent learns to make decisions by receiving rewards or penalties based on its actions.

Edge AI:
AI processing that occurs on local devices (like smartphones or IoT devices) rather than in centralized cloud servers, enabling faster decision-making and reduced latency.

AI Integration:
The process of incorporating AI technologies into existing business systems and workflows to enhance functionality and efficiency.

About the Author

Business Growth Accelerator

Ford Saeks is a Hall of Fame Keynote Speaker and Business Growth Accelerator with over 25 years of experience driving success for companies ranging from start-ups to Fortune 500 corporations. As the President and CEO of Prime Concepts Group, Inc., Ford has helped businesses generate over $1 billion in sales by transforming marketing strategies, improving customer loyalty, and fostering innovation that delivers results.

Trusted Advisor to Global Brands

Having founded over ten successful companies, authored five best-selling business books, and secured three U.S. patents, Ford's influence in the business world is undeniable. His ability to adapt strategies to meet the demands of today's fast-paced economy has earned him numerous industry accolades and positioned him as a sought-after consultant to top-tier companies worldwide.

Hall of Fame Keynote Speaker

A dynamic and engaging speaker, Ford delivers actionable insights that inspire and empower teams to elevate their performance and accelerate business growth. He has shared the stage with industry giants and addressed audiences worldwide, from small startups to billion-dollar brands. His keynotes and virtual training solutions are designed to bridge the gap between strategy and execution, helping organizations implement practical solutions that drive measurable success.

Your Partner in Growth

Ford's expertise spans a wide array of industries, and his dynamic approach has made him a go-to resource for organizations seeking to accelerate their growth. Whether through his keynote speeches, consulting services, or thought leadership, Ford Saeks is dedicated to helping businesses achieve success in today's ever-evolving marketplace.

www.ingramcontent.com/pod-product-compliance
Lightning Source LLC
Chambersburg PA
CBHW071545200326
41519CB00021BB/6612